PLASTICS
TECHNOLOGY:
theory,
design, and
manufacture

PLASTICS TECHNOLOGY: theory, design, and manufacture

William J. Patton

Red River Community College
Manitoba, Canada

RESTON PUBLISHING COMPANY, INC.
Reston, Virginia
A Prentice-Hall Company

Library of Congress Cataloging in Publication Data

Patton, W J
  Plastics technology.

  1. Plastics.   I. Title.
TP1120.P35   668.4   75-17910
ISBN 0-87909-635-7

10 9 8 7 6 5 4 3 2 1

Printed in the United States of America.

contents

PRACTICE

preface

In the use and application of plastics we all suffer from a handicap: metals were in use long before plastics. We are familiar with metals, at ease with them, and we tend to view plastics as a variant of metals. Metals are hard, stiff, and elastic; they corrode; and most important, any metal behaves in principle like any other metal. Anyone familiar with steel, aluminum, and copper would not be out of his element if he had to use rhenium, hafnium, or another unfamiliar metal.

The plastics have none of these characteristics, and our instinct to think in terms of metals technology tends to lead us astray. Unlike the metals, one plastic does not necessarily behave like any other plastic. Many plastics are not even plastic. Indeed, different types of the same plastic may be totally unlike: solid urea-formaldehyde is a hard plastic while urea-formaldehyde foam is the softest plastic. Polycarbonate is widely known as a miraculously tough material—but it is not tough if you cut a notch in it or wipe it with a solvent.

Because of this tendency to think in terms of metals, the early years after the introduction of plastics were filled with optimistic expectation and sorry failures in the use of plastics in consumer and industrial goods. Even today these materials are not sufficiently understood even by technical personnel. They are exciting materials to use, but they are deceptive in ways that no metals are, and the skilled practitioner must use both imagination and caution in working with them.

The theme of this book, therefore, is the understanding of the basic nature of the plastics in general and the peculiar "personality" of each of these materials. Since their characteristics can be altered by their manufacturing processes, a complete understanding of their properties cannot be had without defining that manufacturing process also; the influence of the process on the material must also be discussed. Only an intimate mixture of theory and practice gives a sure management of the plastics. Both theory and suggestions for practice are included in this book.

WILLIAM J. PATTON

THEORY

1

metals
and
plastics

Plastics technology is a development of the twentieth century. The continuous development of new plastics materials has led to new product developments, as well as the substitution of plastics for wood and metals in standard products. Sometimes these developments lead to the popular impression that almost anything can be done with plastics. This, of course, is a false impression. It is as important to know what a material cannot do as to know what it can, and this book will discuss both the advantages and the shortcomings of the plastics. Plastics must compete with other materials, and when plastics are less well adapted to the application, they are rejected for more favored materials. Therefore, to understand how plastics are successfully and unsuccessfully applied, some observations must be made about their competitive materials.

[1.1.]
the solid materials

Most of the engineering materials are solids. There are three broad classes of solid materials:

1.   Metals

2.   Ceramics

3.   Organics

All structural or load-bearing solids are included in these three groups.

A *metal* is an elemental substance, that is, not a chemical compound, which readily conducts both heat and electric current. All the 80 or so metals share these two characteristics, but almost no other characteristics,

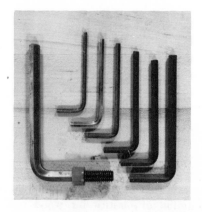

FIGURE 1.1 — This set of Allen wrenches must be made of tool steel, a metal, rather than plastics. The metals provide the required properties of hardness, stiffness, wear resistance, and dimensional stability.

though most metals are hard, lustrous, and stiff. Most metals also are ductile, that is, easily formed or stretched while cold. Among the solid materials, metals for the last hundred years have been the dominant materials because of their broad spectrum of desirable properties:

1.   Hardness

2.   Strength

3. Dimensional stability

4. Rigidity

5. Formability

6. Machinability

7. Weldability

8. Conductivity for heat and electricity

9. Heat and abrasion resistance

Ceramic materials are rock and clay minerals. Familiar examples are sand, glass, brick, cement, concrete, gypsum, plaster, grinding wheels, terrazzo floors, fiber glass, and spark plug bodies, as well as the ores from which metals are extracted. The ceramics have the following properties:

1. Low cost

2. Hardness

3. Low tensile strength

4. High compressive strength

5. Corrosion resistance

FIGURE 1.2 — The plastics cannot provide the properties required in the brick and stone (ceramic materials) of which this armory is built. These properties include high compressive strength, weatherability, hardness, and damage resistance.

6. Electrical insulation

7. Heat and abrasion resistance

The *organic materials* are natural or synthetic materials based chemically on carbon, and include the following:

plastics	wood
rubbers	paper
paints	asphalt
adhesives	fuels
textiles	lubricants
dyes	refrigerants
explosives	foods

The structural organic materials, which include the rubbers, plastics, wood,
and paper, have the following characteristics:

1. Light weight

2. Low conductivity for heat and electric current

3. Fire hazard in some cases

4. Low hardness

5. Lower strength than metals and ceramics

6. Ductility (with some exceptions)

7. Poor dimensional stability

8. Poor resistance to heat

In prehistoric times, the human race had little technology, and therefore
was without metallic or organic materials. Ceramic materials were available
and in most areas wood also. These prehistoric ages have been termed Stone
Ages. Later the human race learned to extract the easier metals from ceramic
ores, and the historic period following such developments is referred to as
the Metals Age or Iron Age. Actually, metals were used only in very limited
amounts until the nineteenth century. The outstanding technical landmark in
all our history was the discovery in the 1860's of two methods of making
cheap steel. The Age of Steel, the dominant material of our civilization, is
thus only a little more than a hundred years old.

In the 1860's there was another invention that was little noted, but was
the forerunner of a second revolution in materials technology. This was the

invention of the first plastic, cellulose nitrate. Although this was a poor plastic in general and a serious fire hazard, it was the first of the synthetic organic solids. It was followed by phenol-formaldehyde (Bakelite) about 1905. Thereafter the pace of development of plastics accelerated. Although currently the weight of plastics consumed annually is only about a fifth that of steels, already the number of cubic inches of plastics consumed exceeds that of steels. Computer projections indicate that by the year 2000 the consumption of plastics will exceed the consumption of steels, and we shall have entered the Plastics Age. The Steel Age then is expected to last only about 150 years, a comparatively brief interval.

[1.2.]
the selection of materials

The selection of a material for a specific application is almost always a thorough, lengthy, and expensive investigation. Almost always more than one material is a possible candidate, and the final selection is a compromise that weighs the relative advantages and disadvantages of all candidates. Prototypes (models) must finally be mocked up in order to test the selected material, and this prototyping provides final insights into the suitability of the material.

The varied requirements demanded of a material may be reduced to three broad demands:

1.   Service requirements

2.   Converting (manufacturing) requirements

3.   Economic or market requirements

The service requirements are, of course, paramount. The material must meet service demands. These may include dimensional stability, corrosion resistance, toughness, adequate strength, scratch resistance, hardness, heat resistance, or any other in an extremely broad range of possible requirements. In earlier decades, when the quality of plastics materials was doubtful and knowledge of these materials was limited, their ability to meet service requirements was regarded with unusual optimism. As a result, vinyl building materials were degraded by sunlight, and cellulose nitrate photographic film set fire to hospitals and movie houses. Nowadays new plastics materials are tested with a more sober professionalism, and no assumptions are made.

Fabricating requirements have tended to grow in importance, and are especially important for plastics. Materials must be easily formable, and it must be possible to join the material to other materials that comprise the finished product. The grade of polyethylene or other plastic best suited to the extrusion process is not the grade that is suited to injection molding. Suppliers, therefore, must know what forming process will be used if they are to supply a suitable formulation of the material.

Finally, there are economic and market requirements. If the product or the raw material is too expensive, it cannot be sold against the competition of cheaper materials. A less tangible requirement is that the product must sell; that is, buyers must have the urge to buy it. They will not do so if they do not like the "feel," the texture, the weight, the balance, the shape, the color, or any other of these more elusive market factors. Consider, for example, that imitation wood moldings are now made of foamed polyvinyl chloride. Why should plastic moldings have to look like wood? For one reason only: Buyers expect them to look like wood. Such a product requirement is not a technical requirement, yet all technical efforts are wasted if such market demands are not satisfied. Marketing considerations are of vital concern to the plastics practitioner, whether he wishes to acknowledge them or not.

[1.3.]
steel and plastics

Technical trends indicate that the dominant materials of the future will be the polymers, not metals. Yet it is impossible to use and shape plastics without employing metals. Plastic products are shaped in dies and molds, and the only suitable mold materials for most plastics manufacturing are steels. Steels, therefore, are an integral part of plastics technology. The study of steels is a major study, and the following remarks lay down only basic approaches to the steel materials.

Alloys are blends of two or more metals. For example, brass is an alloy of copper and zinc. Alloys in metals correspond to copolymers in plastics, a copolymer being a combination of two or more compounds in a polymer, as, for example, ABS (acrylonitrile-butadiene-styrene) or styrene-butadiene rubber. A copolymer or an alloy has a combination of properties that differ from those of the individual elements of the combination.

Steels are alloys of iron, using carbon and often other elements. Few steels contain more than 1 percent carbon, and such high-carbon steels are of little interest to plastics technology. Most steels are low-carbon or "mild" steels, with about 0.2 percent carbon.

Steels are designated according to their carbon contents in the following classification:

DESIGNATION	CARBON CONTENT	AREA OF USE
1. Low-carbon steels	0.03–0.30% C	sheet, plate, structural
2. Medium-carbon steels	0.35–0.55% C	machine parts
3. High-carbon steels	0.60–1.5% C	tools and molds
4. Cast irons	over 2% C	castings

The medium-carbon steels are used for shafts, gears, pins, and other components of machinery, including plastics machinery. High-carbon steels are employed in cutters, knives, shear blades, files, springs, hammers, dies, punches, and molds. Higher carbon gives higher hardness and improved wear resistance.

By a strange coincidence carbon, the fundamental element in polymer materials, is also the essential ingredient in the materials of the present Steel Age. However, plain carbon steels without additional alloying elements do not have sufficient strength, corrosion resistance, or heat resistance for the manufacture of plastics. Polycarbonate, for example, must be injection-molded at temperatures above 500° F and pressures exceeding 20,000 psi. A plain carbon steel, even if high-carbon, cannot sustain such service conditions. A mold steel for such service must contain about 5 percent chromium.

[1.4.]
some remarks for guidance

Everyone is familiar with the general appearance and characteristics of the metals such as aluminum, steels, and stainless steels. If shown an unfamiliar metal such as tungsten or molybdenum, he would not find it unusual; it would resemble the metals with which he is familiar.

Not so with plastics. Plastics do not resemble each other either in appearance or characteristics. Polyethylene is quite unlike phenol-formalde-hyde, and polymethyl methacrylate (Plexiglas) is unlike either of these. You cannot guess the appearance and characteristics of a plastic you have not seen, felt, flexed, scratched, and otherwise examined. Polyethylene, chloro-sulfonated polyethylene, and foamed polyethylene have the same generic name "polyethylene," but these are all unlike materials. Solid urea-formalde-hyde is one of the hardest plastics; foamed urea-formaldehyde is the softest of all plastics.

Consequently, the plastics practitioner or student has a minor difficulty. It will not do simply to read about plastics in a book like this one. You must have samples of the actual plastic so that you obtain some definite concepts of what the material is and how it behaves.

The enormous diversity of materials offered by the plastics makes them exciting materials to work with. It means, in effect, that you can do virtually anything with plastics (though you may not be able to do it economically). Your limitations are not the materials, but your own imagination, creativity, and ability to find solutions to problems. If the characteristics of any plastic are not what you need, then the plastic material can be foamed, filled, reinforced, or laminated to provide entirely new materials of different characteristics. This book will suggest many inventions and design problems to try. Try them. It is not important if your solutions are rather poor—remember that creativity, like any other mental ability, can be developed.

QUESTIONS

1. What basic characteristics of the metallic materials are lacking in the plastics?

2. What basic characteristics of ceramic materials are lacking in the plastics?

3. What advantages do the plastics offer over metals and ceramics?

4. What are the basic deficiencies of the plastics of the present moment?

5. What function does carbon serve in steels?

2

mechanical properties of plastics

[2.1.]
invention and design
in plastics

Inventing and designing are problem-solving processes that begin with an idea. In the development of the initial idea, numerous difficulties and problems arise. Some of these problems can be solved by a systematic attack, while others are solved only by a persistent search for acceptable solutions.

Many inventors have spent a thousand dollars to patent an idea, but without working through the manufacturing and marketing details. As a result, no one is interested in the invention. Any product can, no doubt, be made "of any suitable plastic," but if a suitable plastic has not been selected by the inventor, no company will pay him royalties for the invention, since the search for the right plastic may cost more than the patent costs. The injection-molding die to make the product may cost $30,000, and that die must be designed to suit a specific plastic material.

Other design considerations are more elusive. A change in color, in texture, in shape, in product name, in marketing method, in price, or in discounts may convert an unsuccessful product into a success. These more elusive considerations for marketing and manufacturing are often overlooked by inventors and patentors. To cite an obvious example, very bulky products cannot be sold to chain stores because the product occupies too much shelf space; however, a bulky product can be sold by direct mail methods.

This chapter will discuss some of the considerations for design in plastics. It is not possible, of course, to discuss the whole gamut of design problems from mold making to packaging and shipment of the product, but only the basic technical concepts that apply to all plastic products. But if matters such as color or product name or packaging are not mentioned, it does not follow that these are lesser matters of design. All aspects of design are of equal importance; failure in any one aspect may doom the product. Technical personnel are not usually responsible for marketing, yet technical men who are indifferent to marketing problems are difficult (or impossible) to work with and are capable of dreadful errors in design.

To assist in understanding the process of invention and design, suppose for the purpose of this chapter that we invent an unbreakable plastic hockey stick. A successful unbreakable plastic hockey stick of professional quality could be sold by the millions. A drawing of a wood stick is given in Fig. 8.6. Suppose we make this stick of two plastic materials. The handle will be a foamed high-density high-impact PVC, formed by extrusion. The blade of the stick will be injection-molded from clear polycarbonate, a very tough and abrasion-resistant thermoplastic. The handle and the blade will be joined by that strong and reliable adhesive, epoxy, which will join anything to anything.

In Fig. 8.6 we have the shape, and all the materials have been selected. The cost looks competitive with a wooden stick. We have apparently produced an invention worth a few hundred thousand dollars.

The basic idea is excellent. Unfortunately, the inventor's difficulties are about to begin.

First, try any available epoxy adhesive on polycarbonate. It probably will *not* bond to polycarbonate. Epoxy is indeed a remarkable adhesive, but perhaps not for this application. At this stage we do not have a suitable adhesive, and therefore we do not have a successful invention either.

We cannot here make note of every weakness of this invention—and it has very many indeed—but another may be noted in passing. The blade measures $\frac{5}{16}$ in. thick at the bottom and much thicker toward the handle, according to Fig. 8.6. It just is not practical to injection-mold this thickness of polycarbonate (especially in clear polycarbonate, which will show any air bubbles within the injection-molded piece. There will be such bubbles.)

Another problem. Load the blade of the wooden hockey stick and a polycarbonate blade ¼ in. thick in flexure as shown in Fig. 2.1. The following tabulation shows the results:

LOAD	DEFLECTION OF PC	DEFLECTION OF WOOD
25 lb	0.23 in.	0.06 in.
45	0.42 in.	0.09 in.
85	0.73 in.	0.30 in.
115	—	0.62 in.

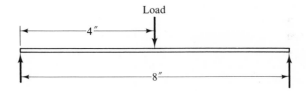

FIGURE 2.1 — Symmetrical loading of a beam flexure.

The PC blade bends very much more than the wood blade. No one will use a plastic hockey stick if it does not give the same action as a wood stick.

This matter of stiffness is the most important limitation of the plastic materials, and therefore it is the characteristic that this chapter should discuss first.

Although these opening remarks have emphasized the problems of invention and design, there is no problem or difficulty that is without a solution. The good designer will find those solutions, though they will require time. A good invention will usually require two years or more of this sort of problem solving.

[2.2.]

stress and strain

The strength of the material is the critical concern for those materials which must carry loads and forces. Not strength, but ductility, may be required of materials in other applications. Springs, vehicle tires, roofing materials, and caulking compounds must be able to "give" or deform. The ability to deform is usually called ductility. Our plastic hockey stick probably has the required strength for its applications, but has too much ductility, both in the blade and in the handle.

While the words "stress" and "strain" are often confused in ordinary conversation, each has a specific meaning. For explanation, consider the case of a square bar of nylon $\frac{1}{2} \times \frac{1}{2}$ in. in cross section and 100 in. long, supporting a tensile pull of 1000 lb. See Fig. 2.2. The stress in this bar is the load divided by the cross-sectional area supporting the load.

$$\text{Stress} = \frac{\text{load}}{\text{area}} = \frac{1000}{\frac{1}{2} \times \frac{1}{2}} = 4000 \text{ lb/sq in.}$$

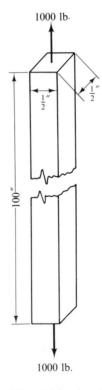

FIGURE 2.2 — Nylon bar in tension.

There are only three possible types of stresses: tension, compression, and shear. These are illustrated in Fig. 2.3. A shear force or stress tends to cut through the material, as occurs in the punching of sheet or plate, or it may be a twisting force such as will twist the head off a bolt. The strength

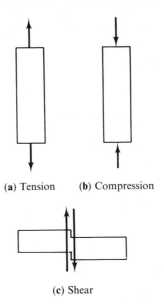

(a) Tension   **(b)** Compression

(c) Shear

FIGURE 2.3 — Three types of stresses.

of materials is lower in shear than in tension, and in the case of most materials, especially brittle materials, the strength in tension is lower than the compressive strength.

Accompanying any stress, there is always some deformation (change in dimension) of the loaded member. In the present example, suppose that the nylon bar lengthens 1.00 in. under its load of 1000 lb. The amount of this deformation will depend on the length of the bar: If the bar were twice as long, it would lengthen twice as much. Therefore, it is convenient to express this deformation as deformation per inch of length. Deformation per inch is the *strain* or *unit strain*, which in the present example is

$$1.00 \text{ in.}/100 \text{ in.} = 0.01 \text{ in.}/\text{in.}$$

Materials may be strained by conditions other than stress. If the temperature of a material is increased, it will expand. This expansion is a thermal strain.

If the load or stress is removed from a material, two types of behavior are possible:

1.   The material may remain permanently deformed by the stress. This kind of strain behavior, in which the strain is permanent, is called *plasticity*. Wet concrete or damp soil are plastic materials; both retain the impression of a shoe if walked upon. Polyethylene is highly plastic in its behavior.

2. The material may return to its original shape and size. This type of behavior, in which the strain disappears with the stress, is called *elasticity*. Elasticity is a characteristic of rubbers. Plastics show two types of elastic strain: The strain may disappear immediately after the stress is removed, or the strain may dissipate slowly with time, after the stress is removed. A foamed polyurethane, for example, may recover its original length as much as a week after the stress is removed.

Many materials exhibit elasticity under limited conditions of stress, but are plastic under heavy stresses, such as structural steels or nylon. Only very brittle materials such as glass and Bakelite (phenol-formaldehyde) are wholly elastic when loaded to their failure stress. In general, the word "brittle" suggests a material in which strain behavior is limited and therefore elastic. Ductile materials are normally those capable of a considerable amount of plastic deformation. Rubbers are ductile but elastic.

The strain behavior of polymers shows three components:

1. An instantaneous elastic strain, explained by the bending and stretching of the bonds between carbon atoms of the polymer chain.

2. A retarded and recoverable elastic deformation.

3. A plastic strain caused by polymer chains slipping past one another.

[2.3.]
stress-strain diagrams

The stress-strain diagram of a material is a most useful compilation of information about any material, since it discloses in a single diagram such characteristics as ultimate strength, limit of elastic behavior, ductility or brittleness, elasticity, plasticity, and modulus of elasticity.

A stress-strain diagram may be given for compression, tension, or shear. The most commonly used is the tensile stress-strain diagram, though compression tests may be more useful for those materials normally loaded in compression, such as concrete or plastic foam insulations.

A typical sample of material for a tension test specimen is shown in Fig. 2.4. The test length in the middle of the specimen is reduced in width or diameter; the larger end tabs are clamped in the jaws of the testing machine. Two light center-punch marks are made in the test length, exactly

Load

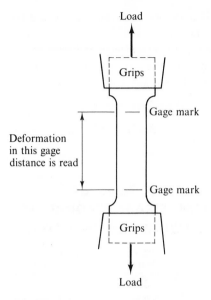

FIGURE 2.4 — Tension test specimen.

2.000 in. apart. The material between these gage marks is the material analyzed under test, and the 2.000-in. length is termed the gage length. A suitable strain gage is mounted on the specimen to indicate the strain in the gage length under load. The specimen is then clamped in a testing machine such as the one in Fig. 2.5. Such a testing machine is actually a highly ac-

FIGURE 2.5 — A universal testing machine for tension and compression testing. The hydraulic loading press is on the left; the load-indicating dial is on the right with flow control valves below the dial.

curate, instrumented hydraulic press that can load materials in tension, compression, or shear. The large dial indicates the load in pounds that is applied to the specimen, while the strain gage indicates the elongation in the 2-in. gage length. For any load, the stress is found by dividing the cross-sectional area of the gage length into the load, and the unit strain by dividing the elongation by 2 (the gage length).

Table 2.1 shows the results of a tensile test on a specimen of mylar (polyester) film measuring 0.006 × 1.00 in. in cross section, with a gage length of 2.000 in.

Table 2.1
STRESS-STRAIN READINGS FOR A
MYLAR FILM IN TENSION

LOAD (LB)	STRESS (PSI)	ELONGATION IN 2 IN.	STRAIN (IN./IN.)
12	2,000	0.014	0.007
23.4	3,900	0.020	0.010
36	6,000	0.030	0.015
67.2	11,200	0.060	0.030
79.2	13,200	0.126	0.063
84.6	14,100	0.216	0.108
90.6	15,100	0.246	0.123
97.8	16,300	0.310	0.155
105.6	17,600	0.380	0.190
110.4	18,400	0.440	0.220
117.6	19,600	0.516	0.258
132	22,000	0.754	0.377
134.4	22,400	0.832	0.416 failure

The graph of stress versus strain is given in Fig. 2.6. The ultimate tensile stress for the specimen is 22,400 psi, which is a high value for a plastic. The maximum elongation is 41.6 percent in 2 in. The strain is linear up to 10,000 psi, this stress being the apparent limit of elastic action or elastic limit. The remaining nonlinear portion of the graph is presumably plastic deformation. Brittle materials do not behave plastically and have maximum strains of less than 10 percent. Since the maximum strain is 41.6 percent for this film, it must be a tough material.

The results for a compression test are shown in Fig. 2.7. The material is a rigid foamed polyurethane with a density of 3.9 lbs. per cubic foot. On being loaded, the material first strains linearly like the mylar film. Note that many plastics and rubbers do not strain linearly at low loadings. Polyethylene

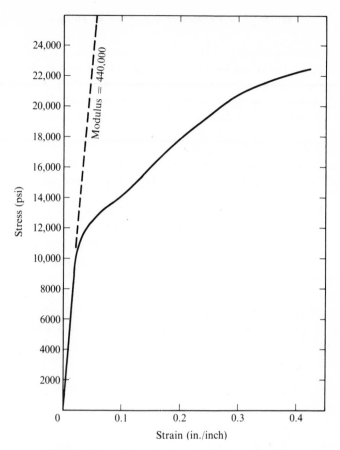

FIGURE 2.6 — Stress-strain curve for a mylar film.

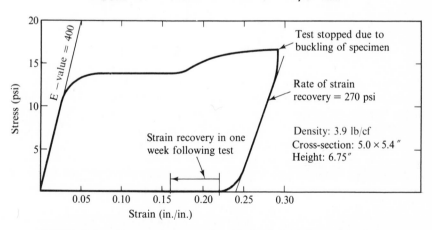

FIGURE 2.7 — Compression test of a block of rigid foamed insulating polyurethane—3.9 lb/cf. Almost one-half of the total strain was recovered in one week.

is one that does not. This linear strain in the polyurethane insulation is elastic. At about 12 psi the strain becomes nonlinear. This is usually an indication of plastic behavior, since the material deforms continually at a constant load. Actually, in this region the material is still in its elastic region, since this strain is recoverable on unloading. At about 18 percent strain the material takes up more load. The loading was stopped at about 17 psi because of buckling of the block of urethane.

On being unloaded, the material recovers elastically at somewhat the same rate shown during the initial loading. This elastic recovery reduces the strain from the maximum of 29 to 22 percent. The curved portion of the strain recovery line near zero stress is the recovery during 15 minutes after the test. After one week, the permanent strain falls to 16 percent. Thus approximately half the maximum strain is elastic and half plastic. Note that the elastic strain recovery is not instantaneous with removal of the load but is delayed in time. Strain is not proportional to stress except in the case of a few materials, usually metals, at low stress levels and slow loading.

A material such as this polyurethane foam, with a maximum elongation of about 30 percent, has excellent ductility. It could extend or contract to accommodate movements of adjacent materials. On the other hand, this material is unsuited to the support of large stresses; it deforms excessively at stresses that are very small. In the whole range of materials, some must be strong and stiff to support loads (polyester), whereas others must be flexible to accommodate movements (foamed polyurethane).

[2.4.]
modulus of elasticity

Most metals, some rubbers, and the elastic plastics such as Bakelite (phenol-formaldehyde), epoxy, and reinforced plastics, exhibit linear elastic behavior below the elastic limit. If this initial portion of the stress-strain graph is linear, that is, if strain is proportional to stress, then the material has a modulus of elasticity. The modulus of elasticity, or Young's modulus, symbol E, is the ratio of stress to strain:

$$E = \frac{\text{increase in stress}}{\text{increase in strain}}$$

More informally, modulus of elasticity is referred to as E-value. The modulus may also be considered as the stress necessary to strain the material elastically to twice its original length, if that were possible. A very rigid material will necessarily have a very high E-value, since a large stress will be accompanied by only a small strain. The slope of the initial curve (E-value) is given in Figs. 2.6 and 2.7. The polyurethane foam has the very low E-value of 400 and the polyester film, an E-value about 1000 times larger.

Table 2.2 suggests that the figure of 1×10^6 separates the high-modulus materials from those of low modulus. Note that all the plastic materials, except reinforced plastics, are low-modulus materials.

Table 2.2
MODULI OF ELASTICITY (PSI)

glass	10×10^6 (varies)	polyethylene	35,000	(varies)
wood	1.5×10^6 (varies)	polyesters	450,000	
concrete	3×10^6 (varies)	acrylics	400,000	(varies)
aluminum	10×10^6	teflon	1,700	
copper	16×10^6	nylons	250,000	(varies)
steel	29×10^6	polycarbonate	300,000	
		polypropylene	200,000	
		polystyrene	500,000	(varies)
		UPVC	400,000	

One of the deficiencies of the plastic hockey stick invented at the opening of this chapter was excessive flex or low modulus. The handle was to be made of high-density foamed PVC, which has a modulus of elasticity of 225,000. This E-value compares most unfavorably with that for hardwoods, which is about 1.6×10^6. The plastic handle will, therefore, flex six times as much as a wood handle under the same force or stress. This is unacceptable. Can the modulus of the plastic hockey stick handle be increased to an acceptable value?

There are at least two methods of increasing the modulus of elasticity of the hockey stick handle and blade: either the composite or the sandwich method. One method would be fiberglass reinforcement. Reinforced plastics may have E-values equal to those of wood. Such reinforced materials are termed *composites*. In the sandwich method, a strip of wood veneer $\frac{1}{32}$ in. thick bonded to both surfaces of a low-modulus plastic would bring the stiffness of handle or blade almost to that of wood. The plastic core cannot yield unless the wood faces yield, and these will strain only about a sixth as much as the plastic, limiting the plastic to a sixth of its normal strain. Alternatively, a RFP (reinforced fiberglass plastic) skin could be laminated to a low-modulus plastic core. Metal skins are used for the same purpose in sandwich construction, and make an even stiffer sandwich.

FIGURE 2.8 — A section through a balsa wood-fiberglass reinforced plastic sandwich. Note that the plastic faces are applied to the end grain of the wood. This arrangement prevents air voids at the wood-plastic bond by allowing air to diffuse into the wood. This balsa wood sandwich is commonly used in large plastic boats and sea-going vessels.

[2.5.]
anisotropy

An isotropic material has identical physical properties in all three dimensions. That is, ultimate tensile stress, maximum elongation, and other properties such as thermal expansion or electrical resistance would be identical, whether determined in the direction of length, width, or thickness of a piece of the material. Isotropic materials, however, are difficult to produce, and further, isotropic properties are not necessarily advantageous. Frequently, optimum properties are required only in one direction, as in the case of the handle of our hockey stick.

When plastic sheet or pipe is extruded, the sheet or pipe must be pulled through the cooling and sizing equipment. This pulling force "orients" the sheet or pipe, resulting in higher strength and lower ductility in the direction of the pull. The finished product, therefore, is anisotropic. Plastic film is usually biaxially oriented, or strengthened in both length and width by stretching. Anisotropy, therefore, is a result of converting operations. Synthetic fibers are given a very high degree of orientation or strengthening in the direction of their length, so that a nylon fiber, for example, will be much stronger in tension than a nylon bar or sheet. On the other hand, a cast sheet should be isotropic.

Anisotropic properties may be disadvantageous for some operations, such as vacuum forming, since formability will be different, as between length and width of the sheet.

[2.6.]

creep

Many materials, including all the plastics and rubbers, are subject to creep. Creep is a slow and continuous increase in deformation under a permanent load, and is a plastic or permanent deformation. The creep behavior of a material cannot be determined from the standard short-time tensile or compressive test; creep failure (fracture) occurs at stress levels well below the failure stress given by a standard tensile test. Hence the allowable stress in a plastic material is determined by the amount of creep that can be tolerated. This creep stress limit may be 40 percent of the short-time failure stress or of that order of magnitude, and if the operating temperature of the plastic material is higher than room temperature, then the allowable stress will be lower, because creep proceeds more rapidly at higher temperatures.

Two types of tests are used to disclose creep properties. In the standard creep test, a stress and a temperature (which may be room temperature or higher or lower) are decided upon, and these are fixed for the duration of the test. The strain in the material is measured at intervals of time, usually up to 10,000 hours. The data are plotted in curves of strain vs. time. Figure 2.9 shows creep data for some thermoplastics at 3000 psi and 23° C.

In the stress-relaxation type of creep test, a certain strain is held constant in the material at a certain temperature, and the stress is measured at intervals of time. The stress will fall in the manner indicated by the curve of Fig. 2.10.

Plastic materials can support very high momentary loadings, but must be conservatively rated for sustained loads. The polyacetal of Fig. 2.9 cannot

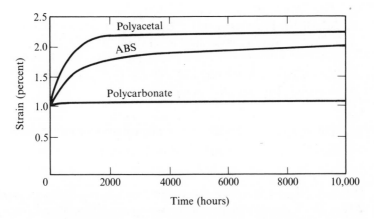

FIGURE 2.9 — Creep curves for three thermoplastics.

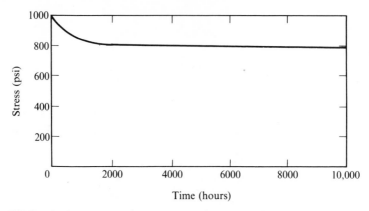

FIGURE 2.10 — Stress-relaxation curve for a polyethylene with an initial applied stress approximating its yield stress.

sustain a continuous stress of 3000 psi, yet it will carry a momentary stress of perhaps 5000 psi, since its ultimate strength is 10,000 psi.

[2.7.]
hardness

Hardness is closely related to strength, stiffness, scratch resistance, wear resistance, and brittleness. The opposite characteristic, softness, is associated with ductility.

Hardness is almost always measured by an indentation test. Some type of indentor is pressed into the specimen; a small indentation indicates high hardness, whereas a large indentation indicates low hardness.

The Brinell hardness tester (Fig. 2.11) uses a 10-mm hardened steel ball as an indentor, using a 500-kg load applied for 30 seconds when plastics are being tested. The theory and operation of the Brinell hardness tester are given in texts on materials science and will not be explained here. Typical Brinell hardness numbers for plastics are these:

polystyrene	25	polyvinyl chloride	20
acrylic	20	polyethylene	2

The most popular indentation hardness tester is the Rockwell machine (Fig. 2.12). There are several Rockwell scales using different indentors and loads. When the Rockwell machine is used to test for the hardness of plastics,

FIGURE 2.11 — The Brinell hardness tester.

FIGURE 2.12 — The Rockwell hardness tester testing a short cylindrical part.

the M and R scales are read. The M scale has a ¼-in. ball for an indentor, preloaded with 10 kilograms, and measures the additional penetration from a load of 100 kg. The hardness is read from the red Rockwell scale of numbers. The M scale is used for the harder plastics. Softer plastics may be tested with the Rockwell R scale, which uses a ½-in. ball and loads of 10 and 60 kg.

The hardness of a plastic as measured by these indentation methods may be influenced by the thickness of the specimen. Typical hardness measurements for plastic materials are the following. Melamine-formaldehyde is the hardest plastic.

melamine-formaldehyde	R_M 120
phenol-formaldehyde	R_M 115
urea-formaldehyde	R_M 115
polyester unreinforced	R_M 70–110
epoxy	R_M 80–110
acrylics	R_M 80–110
polycarbonate	R_M 75
polystyrene	R_M 75
nylon	R_R 110
ABS	R_R 105

The Shore durometer for reading the hardness of plastics and rubbers, Fig. 2.13, is available in several ranges. The type A durometer is used for

FIGURE 2.13 — The Shore durometer for testing the hardness of plastics and rubbers. The durometer is only 2½ in. in diameter.

rubbers and soft plastics and the type D for harder plastics. Figure 2-13 shows the type A instrument. The shape of the two durometer indentors is given in Fig. 2.14. Both have the same diameter but different shapes of point. Both are pressed against the surface of the polymer against the resistance of a spring loading, 822 grams maximum for the type A instrument and 10 lb for type D. The relative movement of the indentor into the specimen is read on the dial of the instrument immediately after the durometer is seated on the specimen. Rubber may creep, and the reading after 15 seconds may be lower than the initial reading. Very hard rubbers, harder than automobile tires, use the D durometer. An automobile tire has a hardness of A70.

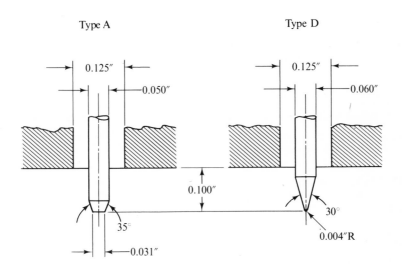

FIGURE 2.14 — Type A and type D indentors for the Shore durometer.

An interesting method of testing for the hardness of plastics is the pencil test. Actually this is a test of scratch resistance. The test uses the whole range of pencil hardnesses from 2B, B, HB, F, H, 2H up to the maximum hardness obtainable, 9H. Each pencil is sharpened to a wedge and dressed with abrasive paper. The pencil is drawn across the surface of the plastic as it is held at right angles to the surface. If the plastic is harder than the pencil carbon, it will not be scratched. If the pencil is harder than the plastic, it will groove the plastic slightly. Pencils are applied in increasing hardness until a hardness is found that first leaves a visible mark in the surface. The hardness of the plastic is reported as the pencil hardness grade.

[2.8.]
specific gravity and weight

Ceramic and most metallic materials are very heavy materials. An outstanding advantage of the plastics is their light weight. This weight is usually reported in handbooks as specific gravity.

Specific gravity is the ratio of the weight of any volume of a material to the weight of the same volume of water:

$$\text{s.g.} = \frac{\text{weight of material}}{\text{weight of water}}$$

Fresh water at 73° F (23° C) weighs 62.4 lb per cubic foot.

EXAMPLE

Polypropylene, the lightest in weight of the solid plastics, has a specific gravity of 0.90 and thus will float on water. What is the weight of a cubic foot of polypropylene?

$$0.90 = \frac{\text{weight of 1 cf PP}}{62.4}$$

Polypropylene weighs about 55 pcf.

Specific weight is the weight of a material in pounds per cubic foot or pounds per cubic inch. *Specific volume* is the number of cubic feet or cubic inches in one pound of material.

Bulk factor is the ratio of the volume of loose and uncompacted molding powder or pellets to the solid volume of the same quantity of material. That is, the bulk volume of powder or pellets includes the volume of the voids between particles. If a loose volume of 4 cubic inches is required to make a solid molding of 2 cubic inches, then the bulk factor is 2:1.

QUESTIONS

1. An injection-molding pressure of 12,000 psi is required to mold the shape of Fig. 2.15. What total pressure must be exerted on the molten plastic?

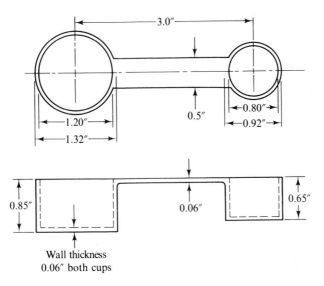

Wall thickness
0.06" both cups

FIGURE 2.15 — An injection-molded dual measuring cup. Molding pressure is applied to the top surface of the cups.

2. The total force of Question 1 is resisted by four circular shafts in tension. If the allowable stress in these shafts is limited to 36,000 psi, what shaft diameter must be used? Choose shaft diameters of even inches or half-inches.

3. Differentiate between: (a) stress and strain; (b) elasticity and plasticity.

4. Explain the usual three components of strain in a plastic material such as polypropylene.

5. Find the modulus of elasticity for the material of Fig. 5.2.

6. Suggest a number of products that require ductility rather than strength.

7. Explain "stiffness" in terms of stress and strain.

8. What is the difference between a composite material and a sandwich construction? Give examples, preferably in sports or leisure equipment.

9. What is the meaning of anisotropy?

10. Give reasons for the anisotropy of wood and plastic film. Why is plywood essentially isotropic?

11. Plastics can support larger stresses for short periods of time than for longer periods. Why?

12. What is meant by stress relaxation under load?

13. In general, abrasion resistance is proportional to hardness in materials. Rubbers are soft, yet abrasion-resistant. Why?

14. Define bulk factor.

15. Differentiate between specific gravity, specific weight, and specific volume.

16. A lightweight polyurethane foam weighs 2 lb/cu ft and a urea-formaldehyde foam weighs 0.7 lb/cu ft. What is the apparent specific gravity of these foams?

17. A certain thermoplastic has a specific gravity of 1.2, and its pellets have a bulk factor of 2.1. What is the weight of a cubic foot of the pellets?

18. The component of Fig. 2.16 is molded in plastic. What is the weight of the component if it is molded from

(a) Polyester, s.g. 1.5?

(b) Polypropylene, s.g. 0.90?

(c) Polystyrene, s.g. 1.04?

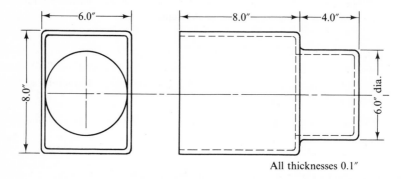

All thicknesses 0.1″

FIGURE 2.16 — Insert minibin.

19. Plastic sailing boat hulls are subject to several types of stresses. Two of these are the pull of the standing rigging (cables) supporting the mast, and slamming pressures due to wave impact. The pull of the rigging is a constant pull at all times; the slamming force acts only for a fraction of a second. Explain why the plastic hull might safely sustain a higher slamming stress than the stress from the rigging.

INVESTIGATIONS

1. Try the pencil hardness test for plastics. What is your opinion of it?

2. Set up a simple bend test to compare the stiffness of plastics with that of woods.

3. Make a simple bend deflection test with a piece of polyurethane or polystyrene foam. Then glue wood veneer to both faces of the foam and try the bend test with the same loads. Can you explain the remarkable improvement in stiffness? If using polystyrene foam, use an adhesive that does not attack the foam.

3

thermal
properties
of plastics

[3.1.]
coefficient of linear expansion

The thermal expansion coefficient is the amount of expansion in a 1-in. or 1-cm length of material as a result of a 1° temperature rise Fahrenheit or Celsius. This expansion is a thermal strain. As a rule of thumb, the plastic materials strain thermally ten times as much as the metals, and since dimensional stability is a highly valued property of materials, this expansion is a serious disadvantage of the polymers. Careful attention must be given to thermal expansion both in converting operations on plastics and in service conditions. Molds for injection molding must usually allow for shrinkage of the plastic on cooling so that the plastic finally cools to the required dimensions. In service, vinyl house siding expands and contracts significantly as compared with wood or aluminum siding.

Consider the expansion in 100 in. of a thermoplastic extrusion and a mild steel bar. The plastic has an expansion coefficient of 0.00005 in./in. $-°$ F and the steel bar 0.000006. The temperature change is 100° F.

$$
\begin{aligned}
\text{Expansion of the plastic} &= 100 \text{ in.} \times 100° \times 0.00005 \\
&= 0.5 \text{ in.} \\
\text{Expansion of the steel} &= 100 \text{ in.} \times 100° \times 0.000006 \\
&= 0.06 \text{ in.}
\end{aligned}
$$

Clearly, metals and plastics cannot be combined unless provision is made for the great difference in thermal expansion between the two types of materials.

In addition to dimensional change from a change in temperature, other types of dimensional instability are possible in plastics:

1. Water-aborbing plastics such as nylon may expand and shrink as they gain or lose water, or even as the relative humidity of the atmosphere changes.

2. Migration or leaching of plasticizer results in slight dimensional change.

3. Traces of unreacted monomer may be delayed in polymerizing. This may result in contraction of a plastic part after it is molded and placed in service. The manufacturer of plastics products has procedures to prevent or control such shrinkage.

4. A plastic product that has been stress-oriented as a result of converting operations may stress-relieve and warp. Again, there are procedures to control such warping.

[3.2.]
thermal conductivity

The rate at which heat can conduct through a material is called its thermal conductivity or K-factor. Metals have very high thermal conductivities and are heat conductors; their K-factors lie between 100 and 3000. All plastics are heat insulators, with K-factors below 1.0. Plastics filled with carbon or aluminum powder will, of course, be more heat-conductive.

Thermal conductivity will be explained here in the units currently used by the construction industry. The construction industry uses the board foot as the unit of volume for board stock, including block and board insulation. A board foot measures 1 foot by 1 foot by 1 inch thick.

The best insulation (lowest thermal conductivity) is provided by foamed polyurethane weighing 2 lb per cubic foot. Consider a board foot of such insulation (Fig. 3.1) with a temperature difference of 1° F across its thickness of 1 in. In one hour there will be a heat flow through this square of insulation of 0.11 Btu. The thermal conductivity of this rigid foam then is 0.11 Btu per hour per square foot per ° F per inch thickness. This is the definition of K-factor or thermal conductivity. If the units are per foot thickness instead of inch thickness, then the K-factor is 0.11/12 Btuh.

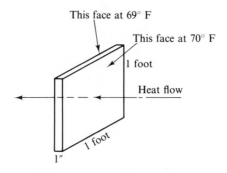

This face at 69° F

This face at 70° F

1 foot

Heat flow

1 foot

1″

FIGURE 3.1 — Thermal conductivity.

The metric units found in handbooks usually give K-factors as calories per second per square centimeter per ° C per centimeter, though such is not the recommended SI unit for thermal conductivity.

The heat flow through a material will be proportional to the area conducting heat and also to the temperature difference between the hot face and the cold face. The heat flow will be inversely proportional to the thickness of the material.

$$\text{Heat flow through a material} = Q \text{ Btuh} = \frac{KA\ \Delta t}{L}$$

where K = K-factor, Btuh/sq ft −° F-inch
A = area of surface, square feet
L = thickness of material, inches
Δt = temperature difference across the material, ° F.

If the thickness is in feet, then K must be per foot thickness.

EXAMPLE

The roof of an industrial building 5000 sq ft in area is foamed with 2 in. of polyurethane with K-factor 0.16. The state is Montana, and temperatures are assumed to be 70° inside and −30° outside. What is design heat loss through this roof?

$$Q = \frac{0.16 \times 5000 \text{ sq ft} \times 100}{2 \text{ in.}}$$
$$= 40,000 \text{ Btuh}$$

Frequently the R-factor or resistance to heat flow is used, especially with construction plastics. The R-factor is the reciprocal of the K-factor. If the K-factor of the polyurethane insulation is 0.16, then its R-factor is 6.2.

The low thermal conductivity of the plastics is an advantage for most of their applications, but not for converting plastics. The forming of plastics requires that they be heated, and it is a slow process, with the risk that the surface of the material may be overheated. Since plastics cannot conduct away very much heat, heat must be supplied in carefully controlled amounts.

[3.3.]
water vapor transmission

There are considerable differences in the rate at which water vapor and other gases can permeate through plastics and rubbers. Polyethylene film is a good barrier for moisture or water vapor, but other gases can permeate through it rather readily. Nylon, on the other hand, is a poor water vapor barrier. Permeability of plastic films is reported in various units, often in grams or cubic centimeters of gas per 100 sq in. per mil thickness of film per 24 hours. Currently the construction industry uses the *perm*.

Many of the activities carried on within buildings greatly increase the water vapor content of the air within the building. Such activities include washing of floors, use of plumbing facilities, cooking, and washing of dishes. Humans and animals contribute water vapor by breathing. Since the air within a building is warm, it can hold a considerably greater amount of water vapor than colder air outside the building. The water vapor pressure, therefore, is higher inside the warm building than in the colder outside air, and consequently vapor will tend to migrate through the walls and roof from inside to outside.

The most severe vapor migration problems occur in cold climates. In the northern states of the Great Plains, minimum temperatures of −30 or −40° F are not unknown. Such temperature differences result in vapor pressure differences of 0.20 to 0.25 psi. These pressure differences cause water vapor to penetrate the walls and roof of a building until it reaches some location within the structure which is at the dew point or temperature of condensation. Here water can collect. In a severe climate it will freeze and destroy roofing and insulating materials.

To prevent such condensation within the building structure, a vapor barrier material is used, such as polyethylene film. This barrier is a layer of material that is impermeable or almost impermeable to water vapor. The vapor barrier must be located on the warm side or interior side of the dew point position in the walls.

The effectiveness of a vapor barrier is rated in *perms*. An effective vapor barrier should have a rating no greater than 0.2 perm. A rating of 1 perm means that 1 sq ft of the barrier is penetrated by 1 grain of water vapor per hour under a pressure differential of 1 in. of mercury. One inch of mercury equals 0.491 or virtually 0.5 psi. One grain is one seven-thousandth of a pound.

A similar problem is presented by a vehicle tire, which must be virtually impermeable to air (oxygen and nitrogen). The most impermeable of the rubbers is butyl rubber, though the carcass of a tire is not made of this rubber. Because of its impermeability to gases, butyl rubber is also used as a roof coating.

[3.4.]
viscosity and melt index

The viscosity and flow characteristics of thermoplastics are a major influence in processing methods and mold design. The word *viscosity* refers to resistance of a liquid to flow. A no. 10 engine lubricating oil has a low viscosity or resistance to flow, whereas a no. 30 oil is "stiffer" or has a higher resistance to flow. It is a familiar experience that the viscosity of fluids is greatly reduced by higher temperatures.

A rigorous explanation of viscosity will not be given here, since this subject is adequately explained in many technical books. Many units for viscosity are in use, but the viscosity of heated thermoplastics is usually given in poises. This unit may be replaced in future by the SI unit, the newton-seconds per square meter. This is not the place to unravel the complexities of these units of viscosity; we shall merely observe that a higher number of poises indicates a "stiffer" or more viscous plastic.

Melt index is the inverse of viscosity, in that a high melt index indicates a low viscosity. The melt index number (MI number) is the number of grams of thermoplastic extruded from a small orifice in 10 minutes at 374° F and 2160 grams load. The melt index indicates the capacity of the resin to flow and fill a mold. The range of MI numbers lies roughly between 0.25 and 25. High molecular weights give low melt indices.

[3.5.]
glass transition temperature

Most thermoplastics are amorphous (noncrystalline); some are partially crystalline.

Suppose a liquid is to be slowly cooled. As the temperature of the liquid falls, so will its specific volume. When the temperature falls to the melting point (freezing point), the liquid solidifies. If the solid phase is crystalline, there will be a sharp drop in specific volume due to solidification, as shown in Fig. 3.2. The solid will continue to contract as its temperature is lowered.

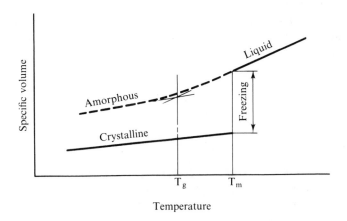

FIGURE 3.2 — Solidification of glassy (amorphous) and crystalline materials.

Now consider a liquid thermoplastic as it is cooled. If it does not crystallize, then there is no sharp melting point. Instead the liquid becomes a supercooled liquid below the (nonexistent) melting point. This supercooled liquid has such a high viscosity that it appears to be a solid material. Glass behaves in this manner: There is a continuous change in viscosity, but no melting point. Such noncrystalline solid materials are called glasses.

On further cooling of the supercooled liquid, the specific volume curve changes its slope in the vicinity of a temperature called T_g, the *glass transition temperature*. Below T_g the material becomes very hard, stiff, and brittle. Above the glass transition temperature the material is rubbery, flexible, and softer. Consider polyethylene and polymethyl methacrylate (Plexiglas) at room temperature, both in the amorphous condition. Polythylene is soft and rubbery, whereas PMMA is hard and brittle. These characteristics indicate that T_g for polyethylene is below room temperature, while T_g for PMMA is above room temperature.

The following tabulation compares the crystalline melting points and glass transition temperatures of some polymers (°C).

	T_g	T_m
Polyethylene	−100	+115
Polyvinyl chloride	+80	+180
PTFE (Teflon)	−115	+325
Natural rubber	−70	+30

Both temperatures are somewhat influenced by the molecular weight (chain length) of the polymer. Higher molecular weight results in higher T_g. Note that of the four resins in the table, only PVC is rigid and glassy at room temperature.

[3.6.]
heat deflection
temperature

The heat deflection temperature of a plastic is useful for assessing load-bearing capacity at an elevated temperature. The method of testing for heat deflection temperature is given in ASTM Specification D648. The sample of the plastic is mounted on supports 4 in. apart and loaded as a beam. A bending stress of either 66 psi or 264 psi is applied at the center of the span. The test is conducted in a bath of oil, with the temperature increased at a constant rate of 2° C per minute. The heat deflection temperature is the temperature at which the sample attains a deflection of 0.010 in.

[3.7.]
flammability and fire hazard

In earlier decades of this century it was thought that the matter of fire hazard was a simple enough one: Does the material burn or does it not? Wood burns. Steel does not. Although these statements about wood and steel are certainly true, they are almost irrelevant to the relative fire risk of the two materials. Compare the cases of fire in two different buildings: one framed of heavy timbers such as glued laminated wood arches, and the other of light steel framing. The light steel frame will collapse after a relatively few minutes of exposure to fire, while it may require a fire of long duration to bring down the timber framing. The problem of fire hazard turns out to be a complicated one.

Fire characteristics of materials are defined in the answers to the following questions:

1.   Is the material flammable?

2.   If flammable, does it ignite readily and spread the flame rapidly?

3.   What are its burning characteristics?

4.   Finally, if it does not burn, how is it influenced by adjacent burning materials?

Some materials may burn quite slowly, but may propagate a flame very rapidly over their surfaces. Thin wood paneling will burn readily, yet a heavy timber post will sustain a fire on its surface until charred and thereafter will smoulder at a remarkably slow rate of burning. Bituminous materials may spread a fire by softening and running down a wall. Steel does not burn, but is catastrophically weakened by the elevated temperatures of a fire. Polyvinyl chloride does not burn, but softens at relatively low temperatures and emits irritating fumes of hydrogen chloride. Other plastics may not burn readily but may emit copious amounts of smoke. Some flammable plastics, such as polyurethane, may be made flame-retardant by incorporating antimony oxide or other suitable additives in the formulation.

There are several tests for comparing the burning rates of materials, including ASTM Specifications D568 and D635. Construction materials are tested for flame spread according to ASTM E84, which is called the tunnel test. A rectangular duct is used, 25 feet long, with the test material as the ceiling of the duct. Burning rates are compared with two arbitrary values:

100 for red oak and zero for cement-asbestos board. Materials with comparative values up to 25 are rated noncombustible.

[3.8.]
a flame spread investigation

Knowledge of the relative flammability of polymer materials is critically important for their specification and use. Also, there is no quicker method of becoming familiar with these materials and their characteristics and identification than to make a flame spread test of them. Such a test should follow more or less closely the procedure of ASTM D635-63.

Test samples of polymer sheet are cut 5 in. long and 0.5 in. wide, and a sharp pencil mark is scribed across each sample at 1 and 4 in. from one end of the sample, as shown in Fig. 3.3. It is preferable to test several samples

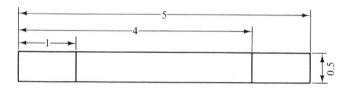

FIGURE 3.3 — Flame spread sample.

of each polymer. The samples are mounted at a 45-deg angle with axis horizontal, as shown in the test procedure of Fig. 3.4. The free end of the sample is ignited with a Bunsen burner or other suitable ignition means. The time for the specimen to burn from one pencil mark to the other is found from a stopwatch or a watch with a sweep second hand, and the average time for all specimens of one type of material is determined. If the sample does not ignite after two attempts at ignition, it is designated "self-extinguishing." If it ignites but does not continue to burn after the igniting flame is removed, it is designated "self-extinguishing."

The usefulness of such a test is greatly extended if it is applied to thin samples of the many woods, papers, and board stock commonly used and then the results are compared with those from the polymers. If possible, obtain a sample of cellulose nitrate for this flame test; this material burns furiously and is therefore unacceptable for many applications.

All pertinent data should be recorded, including thickness of the sample, and for foamed plastics the density in pounds per cubic foot should be known. In recording results note the character of the flame, since a flame

FIGURE 3.4 — A flame spread test on a sample of polystyrene sheet.

test is the most useful method of identifying unknown plastics and rubbers. Note also that many flammable plastics, polystyrene, for example, are available in standard and flame-retardant formulations, and it is important to know which type is being tested.

QUESTIONS

1. A polycarbonate part must measure 5.000 in. after being molded. Total shrinkage of polycarbonate on cooling to room temperature is 0.006 in./in., and the coefficient of expansion of the steel mold is 0.000007 in./in./° F. The mold is heated 200° F above room temperature. What should the mold dimension be to give the part a dimension of 5.000 in. exactly?

2. A polyurethane roof insulation is 100 ft long. If the temperature of this insulation falls 70° and its contraction is not restrained, what is its length at the lower temperature? Coefficient of expansion is 0.00004 in./in. −° F.

3. A cold storage room is maintained at 100° F below the temperature of the other rooms around and above it. The ceiling of this cold room is insulated with 4 in. of polyurethane foam insulation; the area is 800

sq ft, and the K-factor is 0.16 Btuh/sq ft/° F in. thickness. What is the heat flow per hour through this ceiling into the cold storage room?

4. The same cold storage room could have been insulated with less polyurethane foam. Calculate the heat flow per hour if the insulation were 1, 2, and 3 in. thick. What percentage reduction in heat flow does each additional inch produce?

5. What is the purpose of a vapor barrier?

6. What is the meaning of: (a) viscosity? (b) melt index? (c) glass transition temperature?

7. For which of the following materials is the glass transition temperature above room temperature, and how do you know from an examination of the material? (a) Polyethylene; (b) rubber; (c) polystyrene; (d) Teflon.

8. (a) Explain briefly how a heat deflection temperature is obtained. (b) Is this temperature the highest possible temperature at which the plastic may be used?

INVESTIGATIONS

1. Try the flame spread test for a range of plastics.

2. Does polystyrene have a characteristic flame when burned?

3. Can you differentiate polystyrene and ABS plastics by means of a flame test?

4

thermoplastics

[4.1.]
the organics, natural
and synthetic

The phrase "polymer materials" usually means the plastics and the rubbers
(elastomers), although the whole range of polymers includes many other
substances, including wood, foods, and the materials of human and animal
bodies.

Wood, paper, asphalt, coal, petroleum, natural gas, natural rubber,
foods, wool, and cotton, are natural organic materials. All of these natural
materials must be manufactured, converted, and altered in some degree before
being put to use, but they still retain through the manufacturing cycle their

basic characteristics. The use of natural organic materials has tended to decline throughout this century, with the exception of paper, a material that is critical to a civilization based on information processing. The natural organic materials are often superseded by synthetic materials, which are inventions of the chemist. Automobile tires do not use natural rubber; carpets and other textile goods may use artificial fibers, such as nylon and polypropylene. Even paper is often replaced by acetate and mylar film in some applications. Vinyl siding on houses replaces wood. Four percent of the weight of exterior plywood is the synthetic adhesive that bonds the plies together. Much synthetic stone (polyester plastic with crushed stone) replaces natural stone and marble. Finally, even natural rubber (polyisoprene) can be made synthetically, giving us that remarkable material, synthetic natural rubber. The chemist's ability to create new synthetic organics exceeds the capacity of markets to absorb such materials.

Most of the plastics and rubbers to be discussed in this book are *petrochemicals*. Roughly speaking, a petrochemical is an organic material derived from petroleum or natural gas. A petrochemical may contain only carbon and hydrogen, in which case it is termed a hydrocarbon; polyethylene and polypropylene are hydrocarbons. Frequently nitrogen and oxygen or other elements are present also. Sometimes, however, the basic materials for the polymers may be derived from coal. Styrene, with which polystyrene is made, may be obtained either from coal or from petroleum.

Despite the vast range and applications of petrochemicals, they consume only a few percent of coal, petroleum, and natural gas production. The history of petrochemicals probably dates only from 1916, when a method of making propyl alcohol from petroleum was discovered. The synthetic plastics and rubbers joined the list of petrochemicals in limited production in the 1930's, barely in time to serve the needs of World War II. Although it is too involved a question to discuss here, it may be argued that the United States, Great Britain, and Russia could not have won this war without petrochemical research. Developments have come rapidly since that time, and continue at a rapid pace today.

[4.2.]
carbon bonding

In order to understand the characteristics and behavior of the polymer materials, it is necessary to have a general acquaintance with their chemical structure. Fortunately, this chemical knowledge is easy to acquire.

The polymers are built from carbon atoms with associated other elements such as oxygen, hydrogen, nitrogen, chlorine, and fluorine. Carbon provides four chemical bonds to other atoms, or, in the language of the chemist, it has a valence of four (Fig. 4.1). Hydrogen, chlorine, and fluorine, all of which occur in plastic and rubber chemical formulations, have a valence of one, or offer one chemical bond. Oxygen and sulfur have a valence of two.

Formula		Name of chemical	Condition	Use
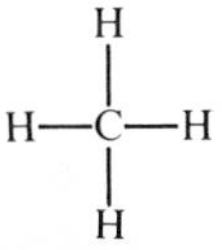 H—C—H (with H above and below)	CH_4	Methane	Gas	Heating fuel
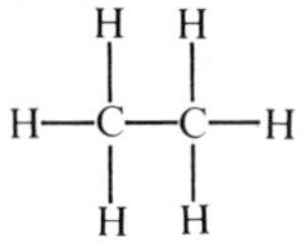 H—C—C—H (with H's)	C_2H_6	Ethane	Gas	Converted into plastics
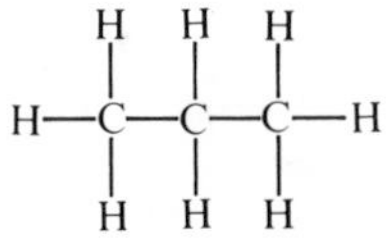 H—C—C—C—H (with H's)	C_3H_8	Propane	Gas	Heating fuel
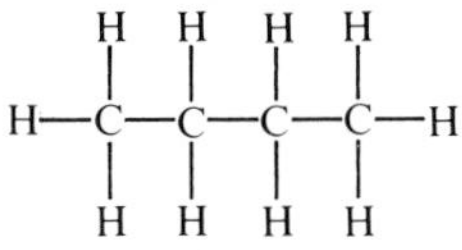 H—C—C—C—C—H (with H's)	C_4H_{10}	Butane	Gas	Heating fuel or converted into rubbers
	C_5H_{12}	Pentane		
	C_6H_{14}	Hexane	Liquid	Converted into engine fuels (gasoline)
	C_7H_{16}	Heptane		
	C_8H_{18}	Octane		
	C_9 .etc.			
	C_{100} .etc.	Asphalts and tars	Solid	Road and roofing asphalts

FIGURE 4.1 — Chemical compounds in natural gas and petroleum.

These chemical bonding arrangements result in compounds such as those of Figs. 4.2 and 4.3. The system of naming these organic compounds is not explained here, nor is it an important matter for the practitioner in plastics.

The quadruple valence of carbon makes possible the formation of long chains of carbon atoms with attached side elements such as chlorine or hydrogen, as in Fig. 4.1 or Fig. 4.7. Many of these long chains are produced by the process of *polymerization*. Polymerization is a process in which two or more molecules of the same substance, such as ethylene, join to give a larger

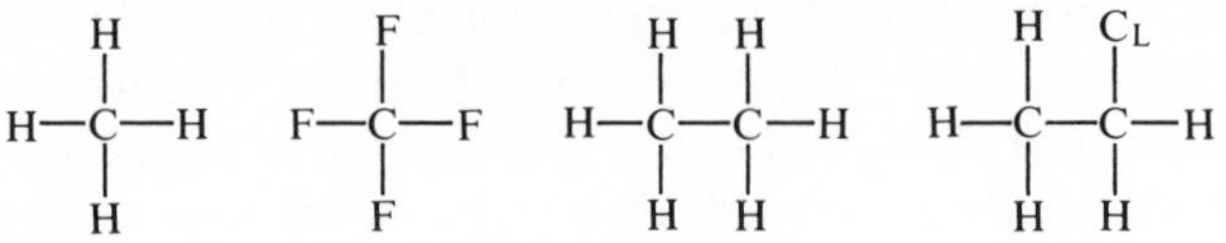

Methane   Tetrafluoromethane   Ethane   Vinyl chloride

CH_4 CF_4 C_2H_6 $C_2H_5C_L$

FIGURE 4.2 — Construction of organic compounds.

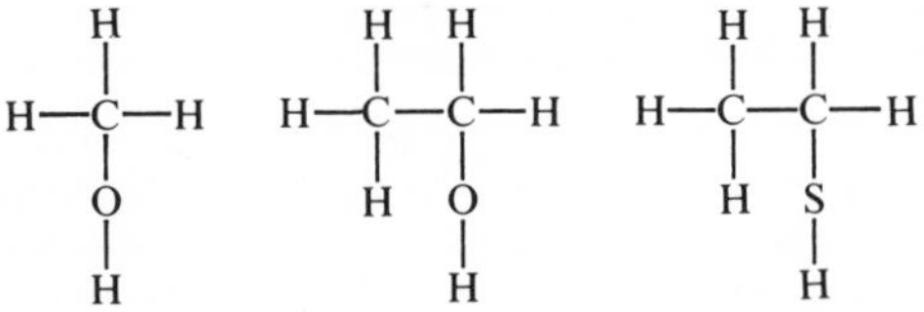

Methyl alcohol   Ethyl alcohol   Ethanethiol

CH_3OH CH_3, CH_2OH CH_3, CH_2SH

FIGURE 4.3 — Organic compounds with oxygen and sulfur.

molecule of the same formula, such as polyethylene, such that the molecular weight of the polymer is some multiple of the unit compound or monomer. Actually, many of the polymerizations to be discussed will not quite conform to this definition of polymerization, but for the present it will suffice.

[4.3.]
hydrocarbon components
of petroleum and
natural gas

The number of hydrocarbon chemicals in petroleum may well be almost infinite. The lighter gaseous and liquid hydrocarbons can be identified and separated from petroleum without difficulty, but most compounds are in the heavier fractions. The individual heavy hydrocarbons cannot be identified, nor can they be separated. Instead, they are divided into groups or "cuts" at the oil refinery. The problem may be understood by an example. The common gasoline hydrocarbon octane, which gave its name to the octane number of gasoline, contains 8 carbon atoms and 18 hydrogen atoms. These

26 atoms can be rearranged to form 17 other hydrocarbons. By reducing the number of hydrogen atoms attached to the 8 carbon atoms, still more hydrocarbons are possible. Twenty-five carbon atoms offer the possibility of about 40 million different hydrocarbons, and 25 is by no means the maximum number of carbon atoms in any petroleum molecule.

Most of the hydrocarbons in petroleum and natural gas (and in coal) can be classified into four groups: paraffins, olefins, naphthenes, and aromatics.

The *paraffins* have the general formula C_nH_{2n+2}. Methane, the chief constituent of natural gas and the lightest of the paraffins, has the formula CH_4. The names of the paraffins end in *-ane*: ethane, hexane, heptane, etc. The paraffins are saturated hydrocarbons, the term "saturated" meaning that they contain the maximum possible number of hydrogen atoms in each case. For this reason they are rather inactive chemically; for example, they are not the best solvents. The smallest or lightest paraffins are shown in Fig. 4.1. Those which form a straight chain without branches are termed normal or n-paraffins; those with branches are isoparaffins.

Paraffins with four or fewer carbon atoms are gases at ordinary temperatures; those with 5 to 15 carbon atoms are liquids; those with over 15 are waxes, asphalts, etc. This general principle is true for plastics: Those with only a relatively few carbon atoms in the polymer chain are liquids, whereas the solid plastics and rubbers are made of very long chains of carbon atoms.

The uses of the lighter paraffins may be briefly summarized as follows:

1. Methane, CH_4 Chief constituent of natural gas

2. Ethane, C_2H_6 Converted into other organic materials such as styrene, polyethylene

3. Propane, C_3H_8 Fuel gas, also converted into other organic materials

4. Butane, C_4H_{10} Liquid fuel, also converted into petrochemicals, especially rubbers

5. Pentane, C_5H_{12}
Hexane, C_6H_{14}
Heptane, C_7H_{16}
Octane, C_8H_{18} Raw materials for gasoline, also other uses

The *olefins*, illustrated in Fig. 4.4, are unsaturated. If two hydrogen atoms can be added to an olefin to saturate it, it is a mono-olefin, C_nH_{2n}. The mono-olefins have names that correspond to those of the paraffins to which they are related, with the ending *-ene* or *-ylene*, thus ethylene, propylene, etc.

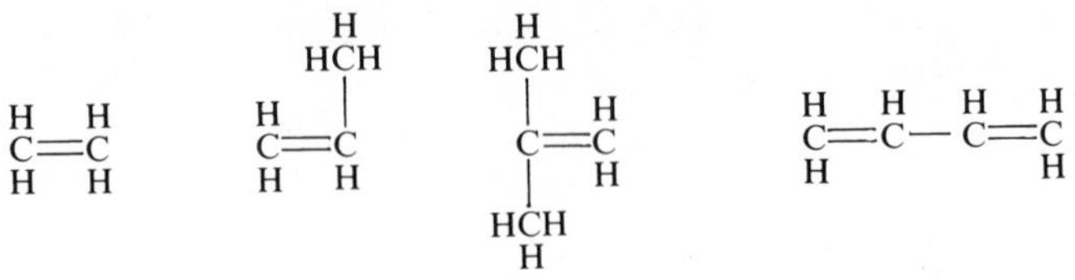

Ethylene Propylene Isobutylene Butadiene

FIGURE 4.4 – Olefins.

If four hydrogen atoms can be added for saturation, the olefin is a diolefin, C_nH_{2n-2}, with a name ending in *-diene* (pronounced *dye-een*), for example, butadiene. Olefins are chemically reactive and serve as the principal raw materials in the manufacture of plastics and rubbers. Actually, olefins are not common in crude petroleum, but are formed in the course of oil refinery operations.

Naphthenes are saturated ring-shaped hydrocarbons with the general formula C_nH_{2n}. The names of the naphthenes begin with *cyclo*. The molecule of cyclohexane is shown in Fig. 4.5.

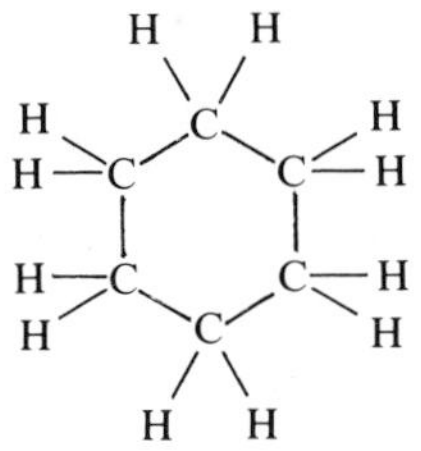

FIGURE 4.5 – Cyclohexane, a naphthene.

Aromatics are found in coal and to a lesser extent in petroleum, especially in California crudes. These are ring-shaped compounds like the naphthenes, but unsaturated. Figure 4.6 shows the structure of benzene, the most familiar of the aromatics. Benzene, toluene, and xylene, often referred to as BTX, are the most important aromatics. The aromatics are chemically active,

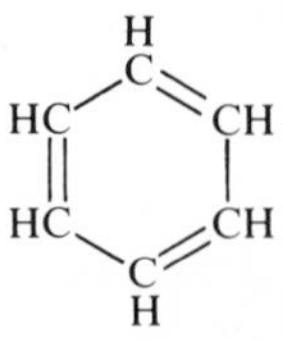

FIGURE 4.6 – Benzene, an aromatic.

and like the olefins are the starting points for the synthesis of a wide range of organic chemicals, including explosives, solvents, dyes, and polystyrene. They are excellent solvents, though they present a fire hazard.

[4.4.]
chemical intermediates

The chemical intermediates are the chemicals such as ethylene, manufactured from the paraffins in petroleum and natural gas, which are then processed further into finished products. The intermediates are the raw materials for almost all rubbers and plastics. Despite their importance, the petrochemical intermediates consume only a few percent of oil and gas production. Intermediates may also be manufactured from the aromatics in coal, which is a source of about one-third of the intermediates used on this continent.

The most important of the intermediates is ethylene. Besides such products as engine antifreeze (ethylene glycol), many plastics and rubbers are produced from ethylene. The chief oil refinery materials used for processing into ethylene are ethane and propane, but butane, naphthas, and gas oils may also be used. In principle, any organic chemical can be made from any fraction of petroleum or natural gas.

Other olefin intermediates are acetylene, which is also a finished chemical product, propylene, butylene, isobutylene, and butadiene. Most of these are used in the production of rubbers, though nylon can be produced from butadiene. Acetylene can be converted into such plastics as polyvinyl chloride and Orlon. Next to the olefins, the most important group of intermediates is the aromatics, chiefly the BTX trio benzene, toluene, and xylene.

The several cellulose plastics, such as cellulose acetate, are not produced from hydrocarbon intermediates, but from wood.

The production of polymers from intermediates is best illustrated by the simplest cases. Consider the first five paraffins:

$$
\begin{array}{lll}
\text{Methane} & CH_4 & \text{or } C_1 \\
\text{Ethane} & C_2H_6 & \text{or } C_2 \\
\text{Propane} & C_3H_8 & \text{or } C_3 \\
\text{Butane} & C_4H_{10} & \text{or } C_4 \\
\text{Pentane} & C_5H_{12} & \text{or } C_5
\end{array}
$$

To make each into its own polymer, it must first be converted to its olefin, methylene, ethylene, propylene, etc. See Fig. 4.7. Note that methylene has two

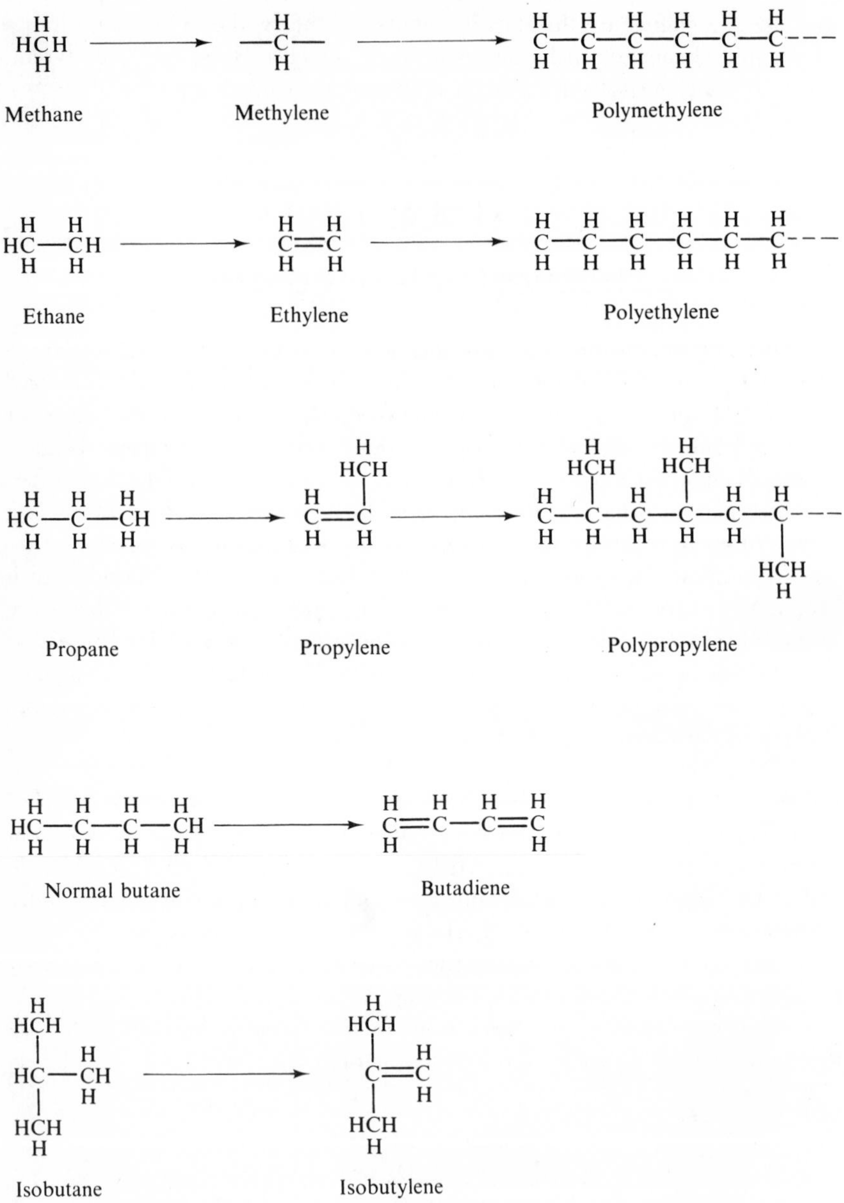

FIGURE 4.7 — Conversion of light paraffins into unsaturated olefin intermediates.

unsatisfied bonds, so that it is an unstable and transitory compound. Butane has two common intermediates, isobutylene and butadiene (pronounced *buta-dye-een*).

These intermediate monomers are polymerized by addition. Thus if 500 or more ethylene monomers polymerize, the result is polyethylene.

Polymethylene has a formula identical to that of polyethylene but a different monomer. Whereas polyethylene is a very flexible and soft plastic, polymethylene is brittle and therefore is not available commercially. Polypropylene is stiffer than polyethylene. Polyisobutylene and polybutadiene are rubbers. Polypentylene is not commercially available.

Although these polymers are usually diagrammed as straight chains, there is some angularity to the linkages between carbon atoms, this angle being 109 deg in the case of polyethylene. The high elongation of many of these materials is explained by the straightening out of these angular linkages. There are also some short side chains attached to the main carbon chain.

[4.5.]
the common thermoplastics

There are two basic types of polymers, *thermoplastic* and *thermosetting*. The thermoplastics are produced by addition polymerization of additional monomers to the chain, and are characterized by their capacity to be repeatedly softened by heating and hardened by cooling. The thermosets are not produced by addition polymerization, and after curing cannot be resoftened. Asphalt is a naturally occurring thermoplastic, softenable by heating. Natural thermosetting polymers are wood, cotton, wool, hair, and feathers. These can be burned, charred, or otherwise damaged by high temperatures, but they cannot be softened. For the present, attention will be directed to the thermoplastic polymers.

The word "resin" occurs frequently when polymers are discussed. Originally it meant a naturally occurring material used in coatings to form a hard and lustrous finish. In a technical sense a resin is an amorphous and high viscosity liquid or solid of high molecular weight which softens on heating. The term now, however, has been extended to include all the polymers used in the plastics industry.

Suppose that we wish to invent a plastic rain gutter for houses (this, of course, has already been invented). The polymers derived from the light petroleum fractions previously mentioned are not suitable for this purpose. The butane polymers are rubbers; polyethylene is too soft, and all these polymers are too combustible for use as a rain gutter on houses. Any chemical material made up of strings of carbon and hydrogen atoms cannot be insured against fire. If no building code or insurance code will allow polyethylene for this purpose, then the ethylene monomer must be altered to make it nonburning. This is done by chlorination, the same method that is used to make a nonburning solvent. By replacing at least one hydrogen atom by chlorine in the monomer, a nonburning thermoplastic material results. Chlorine bonds once, like hydrogen. The modified monomer is called vinyl chloride (Fig. 4.8).

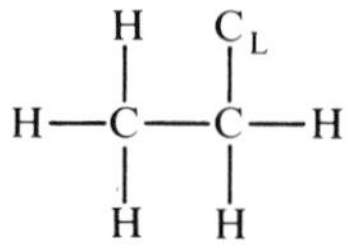

$$H-\underset{\underset{H}{|}}{\overset{\overset{H}{|}}{C}}-\underset{\underset{H}{|}}{\overset{\overset{C_L}{|}}{C}}-H$$

FIGURE 4.8 — Vinyl chloride monomer.

The resulting *polyvinyl chloride* proves to be an excellent choice for a rain gutter. It does not burn, and unlike polyethylene it is a stiff plastic. It is also inexpensive and can be made resistant to the deteriorating effects of ultraviolet radiation in sunlight. A rain-gutter fitting of PVC is shown in Fig. 4.9.

FIGURE 4.9 — An injection-molded polyvinyl chloride tee fitting for an eavestrough. The two nail holes are for attachment to the building. Polyvinyl chloride troughs snap into the two arms of the tee, and the downspout snaps over the bottom leg.

If we substitute two chlorine atoms for two hydrogen atoms in the ethylene molecule, *polyvinylidene chloride* (Fig. 4.10) results. The addition of chlorine atoms can be carried to the maximum of four.

The element fluorine, with a single bond, may also be substituted for hydrogen to produce nonburning thermoplastics. A single fluorine atom in

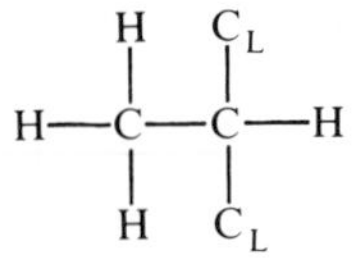

$$
\begin{array}{c}
\quad\;\; \text{H} \quad\; \text{C}_\text{L} \\
\quad\;\; | \qquad | \\
\text{H—C—C—H} \\
\quad\;\; | \qquad | \\
\quad\;\; \text{H} \quad\; \text{C}_\text{L}
\end{array}
$$

FIGURE 4.10 — Vinylidene chloride monomer.

the ethylene monomer results in the *polyvinyl fluoride* of Fig. 4.11. Polyvinyl fluoride is a thermoplastic of outstanding properties, is nonburning, and is used as a surface film to protect other materials from weathering, ultraviolet degradation, corrosion, staining, or damage by scuffing. It is sold under the trade name Tedlar.

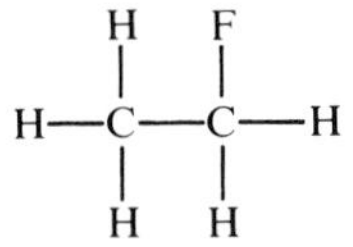

$$
\begin{array}{c}
\quad\;\; \text{H} \quad\; \text{F} \\
\quad\;\; | \qquad | \\
\text{H—C—C—H} \\
\quad\;\; | \qquad | \\
\quad\;\; \text{H} \quad\; \text{H}
\end{array}
$$

FIGURE 4.11 — Vinyl fluoride monomer.

Two fluorine atoms give *polyvinylidene fluoride*, another thermoplastic of superior properties and used as film, like PVF. Four fluorine atoms produce the well-known Teflon, *polytetrafluoroethylene*, with outstanding resistance to corrosion and high temperatures. A great many other monomer possibilities can be designed by using fluorine or chlorine or both.

Clearly, the chemist, or even the reader of this book, can continue to "invent" thermoplastics based on ethylene almost without end. Each formula so easily invented would, of course, have to be manufactured and tested for properties and cost. Many, such as polymethylene, would be rejected for reasons of properties, cost, or processing limitations.

[4.6.]
the polymer molecule

The synthetic thermoplastics are molecules that are chains of 500 or more carbon atoms (the chain may contain atoms other than carbon), the distance between adjacent carbon atoms being closely 1.5×10^{-8} cm. In discussing the length of a polymer molecule we are necessarily referring to an average length, since it is not possible, nor is it necessary, for all molecules to be composed of the same number of monomers. However, some control over the length of the molecule must be possible.

Most monomers are gases. A short polymer chain of low molecular weight will be a liquid. If the molecular weight is large enough, the polymer will be a solid. The entanglement and attraction between such large molecules accounts for their mechanical strength. Higher molecular weight results in increased strength and stiffness, as illustrated in Figs. 4.12 and 4.13. As is usual for any material, increased strength is accompanied by decreased ductility. The increase in mechanical properties levels off as molecular weight becomes exceedingly large.

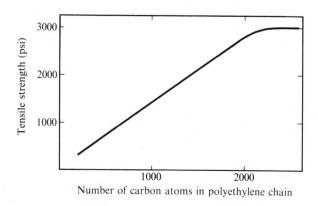

FIGURE 4.12 — Variation of tensile strength with length of carbon chain in a typical polyethylene.

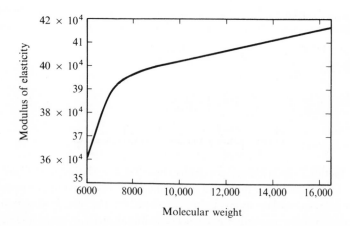

FIGURE 4.13 — Effect of molecular weight on the modulus of elasticity of polyvinyl chloride-acetate.

For polyethylene, ethylene is a gas, a polymer of 6 monomers of ethylene is a liquid, 36 is a grease, 140 is a wax, and 500 is the familiar plastic.

The best mechanical properties, therefore, are provided by the highest possible molecular weight. However, the very largest molecular weights are the most difficult to process in such plastic converting operations as extrusion, injection molding, and blow molding. Therefore, the molecular weight selected is the best compromise between properties and processability. In the case of rubbers, coatings, and thermosets, a low molecular weight is used for ease of processing, and this low molecular weight is later converted into a high molecular weight for best properties. A very high molecular weight is accompanied by a very high melt viscosity. To reduce this viscosity the processing temperature must be elevated, and the resulting excessively high temperatures degrade the plastic and increase the cost of production. At the same time the molecular weight must be adjusted to the process: An extrusion grade of thermoplastic has a higher molecular weight than an injection-molding grade, because the hot extrusion must have sufficient viscosity to support itself when it leaves the die. For a plastic foam, the melt viscosity must be low enough to permit expansion by the foaming gas, but high enough that the thin walls of the foam cells do not rupture during expansion. Hence, like metals, plastics must be formulated to suit the manufacturing process.

To improve the processability of some polymers of high molecular weight, especially polyvinyl chloride, plasticizers may be added to the material. Unplasticized polyvinyl chloride (UPVC) is hard and brittle. When plasticized, it is suitable for garden hose, shower curtains, raincoats, and packaging film, all of which require a rubbery flexibility. The plasticizer, being a liquid, has the effect of lowering the average molecular weight, thus reducing stiffness and increasing flexibility.

Many thermoplastics, though not all, are deteriorated by prolonged exposure to oxygen of the air and ultraviolet radiation in sunlight. Resistance to such deterioration is improved by higher molecular weight for three reasons:

1.   Higher molecular weight means fewer molecules per pound or per unit volume. Fewer molecules means fewer terminal monomers in the carbon chain, and these terminal monomers are necessarily the reactive monomers.

2.   The longer molecule is less mobile.

3.   A high molecular weight, if degraded, will degrade only to a medium molecular weight with acceptable properties instead of to a low molecular weight.

Aging has as one of its effects the reduction of molecular weight.

[4.7.]
crystallinity

The simplest of the polymer structures is polyethylene. In the molecule of polyethylene, the hydrogen atoms lie in planes perpendicular to the plane of the carbon atoms, which have a zigzag arrangement. See Fig. 4.14. Polytetrafluoroethylene (Teflon) is similar to polyethylene, with the four hydrogen atoms of polyethylene replaced by four fluorine atoms. But in PTFE the carbon chain describes a helix, with 14 carbon atoms per turn of the helix.

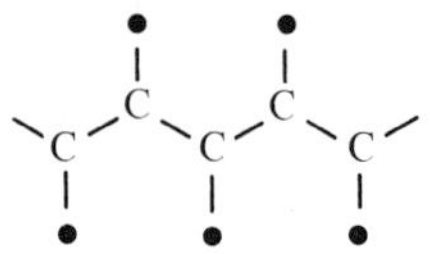

FIGURE 4.14 — Configuration of the polyethylene molecule. Each carbon atom has two hydrogen atoms attached, one above and one below the plane of the paper.

A polymer with the shape of a coil spring, and with side chains or attached methyl groups (CH_3) or aromatic rings, has a shape that is hardly susceptible to the regular arrangements of atoms that characterize a crystal structure. It is not, therefore, surprising that many polymer materials are amorphous or glasses. But often a degree of crystal ordering on a submicroscopic scale is possible. Such ordered regions are called *crystallites.* The whole molecule, because of its great length, may extend through several crystalline and noncrystalline regions, as illustrated in Fig. 4.15. The greater the crystallinity or number of crystallites, the greater the density of the polymer, because of the closer packing of the molecules in the crystalline phase.

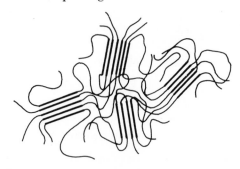

FIGURE 4.15 — The coiling of long polymer chains through several crystalline and noncrystalline regions.

Thus low-density polyethylene may be as much as 55 to 70 percent crystalline, whereas high-density polyethylene may be 75 to 95 percent crystalline.

An advantage of crystallinity is that the polymer product may be used at higher temperatures than are possible with a more amorphous product. Instead of softening gradually with increased temperature, the crystalline polymer tends to exhibit a sharp melting point characteristic of any crystalline material. A more crystalline polymer will have a better resistance to water absorption and to solvent attack. The crystalline regions are closely ordered segments, while the amorphous regions present an open and random arrangement of atoms that can be penetrated by water, solvents, or permeating liquids. High crystallinity makes the use of plasticizers impractical, because the plasticizer cannot penetrate the crystallites. Therefore, plasticizers are not used with crystalline polymers such as polyethylene and nylon. Similarly, the crystalline polymers tend to be impermeable to gases, a characteristic that may be useful in food packaging or in protective coatings. Such impermeability is a disadvantage in polymer fibers, which usually must receive dyes.

Ultimate tensile strength increases with crystallinity, as shown for polyethylene in Fig. 4.16, but the reduced mobility of the molecules due to

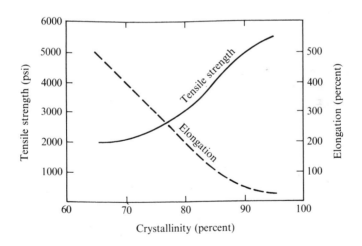

FIGURE 4.16 — Effect of degree of crystallinity on ultimate tensile strength and elongation of polyethylene.

crystallinity reduces ultimate elongation. The modulus of elasticity of polyethylene is greatly influenced by crystallinity, as shown in Fig. 4.17.

The relative crystallinity can be altered by heat treatment in the case of those polymers with the capacity to crystallize. A crystallizable poly-

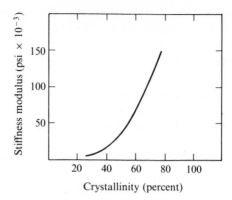

FIGURE 4.17 — Influence of the degree of crystallinity on the stiffness of polyethylene.

ethylene can be given a crystallinity of about 80 percent by slow cooling, or 65 percent by rapid cooling.

Besides the amorphous and crystalline conditions, an intermediate condition, the *oriented* condition, can be produced. If a polymer is drawn into fibers through a die, the molecules and microcrystals of the polymer are aligned in the direction of drawing. The oriented condition gives very high tensile strength levels and impact strength in the direction of orientation. Polymer fibers, being oriented, are considerably stronger than bulk polymers. The production of polyethylene film employs extrusion with air-blowing, as seen in Fig. 12.8; such film is oriented in both the longitudinal and cross direction, with the additional advantage that the film is improved in clarity.

Crystallinity is improved by ordering in the chain of monomers. The molecule of polypropylene is shown in Fig. 4.18. If the methyl unit (CH_3) is attached at random locations above and below the carbon chain as in the first part of the figure, then the polymer cannot crystallize. Such an irregular configuration is called the *atactic* structure. A regular arrangement of methyl groups, as in the second part of the figure, makes crystallinity possible. The isotactic and atactic structures are produced by different catalysts and processing conditions during polymerization. In the isotactic structure all the

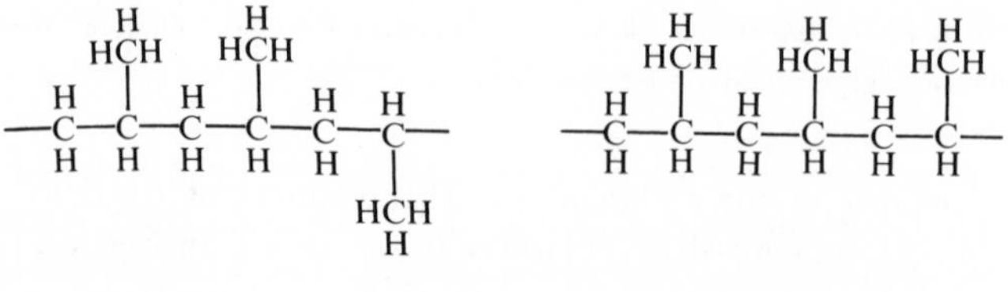

FIGURE 4.18 — Atactic and isotactic polypropylene.

methyl units are located on one side of the carbon chain; they alternate on both sides in the syndiotactic configuration.

[4.8.]
stress-cracking

When a polymer is subjected to a moderate stress and at the same time exposed to a weak solvent or surface-active material for a prolonged period, a brittle type of failure called environmental stress-cracking may occur. The chemical causing the stress-cracking may be a relatively harmless chemical, such as a soap or a detergent. Stress-cracking probably is initiated at the weaker amorphous regions between crystallites, even though the condition is often more commonly found in polymers of high crystallinity. Low-molecular-weight fractions in the polymer contribute also to stress-cracking. The stress need not be an applied stress, but may be a residual (internal) stress resulting from the forming process used to shape the plastic part. A blow-molded bottle contains such locked-in stresses at its corners, and if filled with detergent may perhaps fail by stress-cracking. Polyethylene and polystyrene especially are susceptible to this type of failure.

[4.9.]
standard abbreviations

The following discussion of the characteristics and properties of specific thermoplastics must be limited to those resins in common use or of unusual significance. All these plastic materials are made in a range of formulations; an extrusion grade of thermoplastic, for example, is not quite the same as an injection–molding grade. Therefore, when values for such properties as ultimate tensile strength are reported in the text, the reader should take such values as order-of-magnitude or "ball-park" figures. A polypropylene fiber has a much higher strength than a polypropylene film, and the film has a higher strength than an injection-molded polypropylene article.

Three plastics are soluble in water. These are polyvinyl alcohol, methyl cellulose, and polyethylene oxide. These will not be further discussed.

The American Society for Testing and Materials (ASTM) recommends the following abbreviations. Both thermoplastics and thermosets are included in the list, which is, of course, not a comprehensive listing of the plastics.

ABS	Acrylonitrile-butadiene-styrene
CA	Cellulose acetate
CAB	Cellulose-acetate-butyrate
CN	Cellulose nitrate
EP	Epoxy
MF	Melamine-formaldehyde
PF	Phenol-formaldehyde
PAN	Polyacrylonitirle
PA	Nylon (polyamide)
PC	Polycarbonate
PE	Polypropylene
PETP	Mylar (polyethylene-terephthalate)
PMMA	Acrylic, plexiglas, lucite (polymethyl methacrylate)
PP	Polypropylene
PS	Polystyrene
PTFE	Teflon (polytetrafluoroethylene)
PVAC	Polyvinyl acetate
PVAL	Polyvinyl alcohol
PVB	Polyvinyl butyral
PVC	Polyvinyl chloride
PVF	Polyvinyl fluoride
UF	Urea-formaldehyde
UP	Urethane plastic

[4.10.]
the polyolefins

The polyolefins include the plastics PE and PP and the rubbers polybutadiene and polyisobutylene.

Polyethylene is the most widely used of all the plastics. It was discovered just before World War II, and like so many new materials, its possible uses were unknown at the time of discovery. It was a solution looking for a problem, and that problem turned out to be the insulation on high-frequency radar cables, for which purpose PE was eminently suited by reason of its superior electrical properties. Thus polyethylene was discovered just in time to make radar feasible, and radar made it possible for Fighter Command of the British Royal Air Force barely to win the crucial battle of World War II, the air battle of Britain in 1940.

Polyethylene is a true plastic in that it is capable of large plastic deformation. Its strength is low, with low elastic modulus, high expansion and mold shrinkage, and creep under stress at room temperature. It is thus not suited to applications requiring dimensional stability. Despite these limitations, PE has developed remarkably large markets because of its low cost, light weight, resistance to chemicals, and pleasant feel. Its uses include electrical insulation, especially for high frequencies, film, pipe, containers and bottles, paper coatings, and injection-molded articles such as kitchenware, laboratory apparatus, and toys.

Polyethylene is produced in three grades: low density, intermediate density, and high density. The specific gravities of these three types are

low density	0.912–0.925
intermediate density	0.925–0.940
high density	0.940–0.965

The degree of crystallinity is proportional to the specific gravity, and the heat resistance improves with specific gravity. The higher densities are less waxy in appearance and to the touch. The higher densities are more crystalline because they have relatively few side branches. Low-density materials are highly branched, and have greater elongation and lower softening temperatures. Resistance to stress-cracking is better in low-density polyethylene. High-density PE has greater mold shrinkage.

All the polyolefins are degraded by ultraviolet radiation. Where resistance to weathering and ultraviolet radiation is a requirement, PE is blended with about 2 percent carbon black.

The largest market for polyethylene is film, most of this film being consumed by the packaging industry. The film is used by the construction industry as a vapor barrier for walls and for concrete while it is setting. Most film is made of low-density material because of the lower cost, though high-density PE is a much better vapor barrier. Polyethylene construction film is available in clear, white, and black colors, and thicknesses of 2, 4, and 6 mils. One mil — 0.001 in.

Polyethylene, like other thermoplastics, may be made to cross-link between molecular chains by means of gamma radiation in very heavy doses. When it is subject to such radiation, hydrogen is liberated and carbon atoms can then cross-link. The polyethylene then becomes infusible (thermosetting) and withstands prolonged aging at temperatures of almost 150° C.

Propylene is a gas boiling at −48° C at atmospheric pressure. Its polymer polypropylene is the lightest of the common plastics, with a specific gravity of 0.90 (methylpentene is even lighter). General physical and electrical properties of polypropylene are similar to those of high-density polyethylene. Polypropylene, however, is harder and has a higher softening point, lower shrinkage, and immunity to stress-cracking.

Polypropylene is readily degraded by ultraviolet radiation and heat. Stabilizers are blended into the resin to protect it against processing temperatures.

Polypropylene film is not as limp as polyethylene film. In addition to film, this plastic is molded into domestic hollow ware, toys, bottles, automotive components such as distributor caps, disposable syringes for medical and veterinary use, battery cases, rope, and carpeting.

[4.11.]
vinyl and related polymers

Vinyl chloride and vinyl fluoride monomers are shown in Figs. 4.8 and 4.11, and a vinylidene monomer in Fig. 4.10. The vinylidene molecule contains two chlorine or fluorine atoms, as compared with one in the vinyl molecule.

The more important of the vinyl polymers are PVC, either plasticized or unplasticized, copolymers of polyvinyl and polyvinylidene chloride (Saran), polyvinyl acetate (used in paints and adhesives), polyvinyl butyral (the adhesive for safety glass), and polyvinyl and polyvinylidene fluoride, both employed only as film.

Rigid or unplasticized polyvinyl chloride (UPVC) is a hard thermoplastic, nonburning, and capable of being formulated for weather resistance. It is now familiar in such uses as pipe and ducting, fume hoods, and such construction products as house siding, window sash, and raingutters and downspouts. Flexible or plasticized PVC contains 15 to 50 percent of plasticizers for softness and flexibility. Uses of the plasticized polymer include garden hose, electrical wire insulation, shower curtains, and clothing fabrics. The plasticized material is not suited to exterior use in building products because of the tendency of the plasticizer to leach out over extended periods of time.

Polyvinylidene chloride is a stiff plastic like PVC. When copolymerized with 30 to 50 percent of vinyl chloride, the resulting copolymer is soft and flexible. Dow Chemical Corporation calls this copolymer Saran. A copolymer is any polymer which contains two or more monomers in its chain.

Saran has a strong tendency to "block" or cling to itself. It is the most impermeable of the transparent films to gases and water vapor.

Polyvinyl fluoride has outstanding tensile strength, ease of maintenance, and resistance to ultraviolet radiation, staining, and abrasion. It is used only as a protective film over wood panels, furniture, upholstery, and other plastics, in thicknesses of 0.5, 1, 1.5, and 2 mils. Polyvinylidene fluoride is used as a metal finish as well as a protective film.

Polyvinyl fluoride has one fluorine atom, while polyvinylidene fluoride has two. Polychlorotrifluoroethylene (CTFE) has three, the fourth atom being chlorine. Increasing numbers of fluorine atoms in the monomer result in greatly increased ability to sustain high temperature. CTFE can be used at temperatures as high as 400° F or 200° C; polytetrafluoroethylene (PTFE), with four fluorine atoms, can be used at 250° C. Such fluorinated polymers also have excellent electrical characteristics, and remarkable resistance to chemical attack.

Polytetrafluoroethylene (Teflon) is a crystalline (over 90 percent crystallinity) thermoplastic, white and opaque, soft and waxy. Though the monomer boils at −106° F, PTFE can be exposed to continuous temperatures of 500° F. With such temperature resistance, the polymer will be difficult to process and form. It has little strength and creeps readily at room temperature. PTFE provides the lowest unlubricated coefficient of friction of any material: about 0.04.

Fluorinated ethylene-propylene (FEP) is the only other carbon chain with fluorine attached atoms only. Its applications are those of CTFE and PTFE, with a temperature rating of 400° F.

Polystyrene, like PVC and PE, is a widely used low-cost thermoplastic. Because of the aromatic ring in the monomer, it does not crystallize. Two general types of PS resins are available: general-purpose and impact grades. Typically the impact type is copolymerized with about 5 percent butadiene, a rubber.

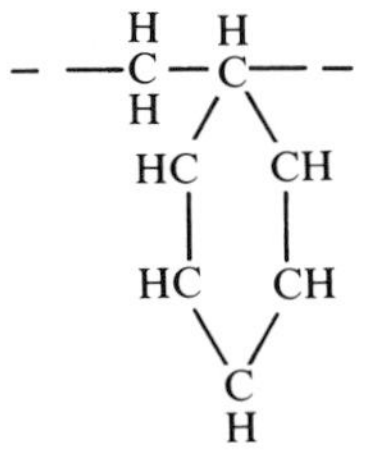

FIGURE 4.19 — Chemical structure of polystyrene.

Polystyrene is a crystal-clear plastic, hard, brittle, and low in impact resistance. It has a brilliant surface. However, it is easily stress-cracked, and is not recommended for holding liquids. Polystyrene tumblers for kitchen use usually fail by stress-cracking as a result of continued exposure to compounds used in dishwashing. Solvent resistance is poor, and the resin is rapidly degraded by sunlight. Its heat resistance also is limited. The monomer formula indicates that polystyrene is combustible, and it can be identified by its characteristic blue flame when burned. (ABS plastic gives an identical flame.)

When polystyrene is bonded with solvent cements, fast evaporating solvents should not be used because they may result in stress-cracking. Methylene chloride, trichlorethylene, toluene, and xylene are usually successful solvent cements for this plastic.

Acrylonitrile-butadiene-styrene (ABS) is a copolymer, characterized by excellent toughness. A great many types of ABS are available, some of these being alloyed with other plastics such as PVC or polycarbonate. ABS is used in DWV pipe (drain, waste, and vent) for plumbing and drains.

[4.12.]
nylon and acrylic

Nylon and acrylic were the first of the "engineering" or high-performance thermoplastics.

Nylons are polyamides produced by the reaction of a diamine with an organic acid. The most commonly used of the nylons is produced by reacting hexamethylenediamine with adipic acid to give hexamethylenedipamide, which is called nylon 6/6, that is, there are six carbon atoms in the amine and six in the acid segments. There are other grades, such as 6/10, 6/12, 13/13, etc.

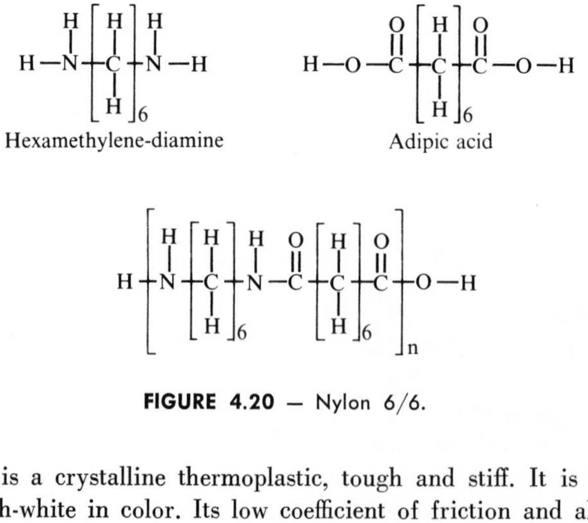

Hexamethylene-diamine   Adipic acid

FIGURE 4.20 — Nylon 6/6.

Nylon is a crystalline thermoplastic, tough and stiff. It is translucent and yellowish-white in color. Its low coefficient of friction and abrasion resistence account for its use as gears, bearings, and cams. Nylon bearings do not require lubrication.

Nylons will absorb moisture, even from the atmosphere. Absorbed moisture acts as a plasticizer, lowering the stiffness, strength, and hardness of the nylon. See Fig. 4.21.

$$
\begin{array}{c}
\text{H} \\
| \\
\text{HCH} \\
\text{H} \quad | \\
- \ -\text{C}-\text{C}- \ - \\
\text{H} \quad | \\
\text{C}=\text{O} \\
| \\
\text{O} \\
| \\
\text{HCH} \\
\text{H}
\end{array}
$$

FIGURE 4.21 — Stress-strain curves for dry and moist nylon at 70° F.

Polymethyl methacrylate, PMMA, is more familiarly known by its trade names, such as Lucite or Plexiglas. This thermoplastic has some outstanding properties. It is crystal clear, transparent (transmittance for light is 92 percent, slightly better than glass), with almost unlimited color possibilities. The cast product has better optical properties and higher strength than extruded sheet. Dimensional stability is superior to that of most thermoplastics.

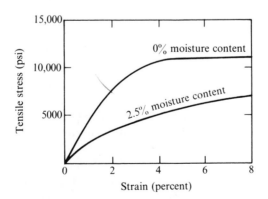

FIGURE 4.22 — Polymethyl methacrylate.

PMMA is familiar in its applications: lighting fixtures, outdoor signs, aircraft glazing, tail lights of automobiles. It is used instead of window glass where vandalism is a problem. The impact resistance of PMMA is better than that of glass, though polycarbonate is far superior to both as a tough glazing material.

[4.13.]
cellulose plastics

Wood is a cellular thermosetting material, and, like the other thermosets, is an elastic material. Only recently has it been possible to make foamed plastics that even approach the properties of wood.

Chemically, wood is a compound polymer composed of three principal polymeric constituents. The wood of a tree consists of long tubular fibers. These fibers are composed of two polymeric carbohydrates called cellulose and hemicellulose. Both are complex glucose compounds. Glucose is a sugar, which explains why fungi, insects, and animals can use wood as a food. These wood fibers are bound together with a third polymer called *lignin*. The fiber bundles run the length of the tree. Based on an oven-dry condition, the constitution of woods is the following:

	HARDWOODS	SOFTWOODS
Cellulose	40–45%	40–45%
Hemicellulose	15–35%	20%
Lignin	17–25%	25–35%

Small amounts of other substances are also present in woods.

It is cellulose that provides the strength in axial tension, toughness, and elasticity of wood. The long-chain molecules of cellulose are in bundles that run helically to form hollow needle-shaped cells or fibers. These fibers range in length from 1 to 3 millimeters, and have thicker walls in hardwoods than in softwoods. The long-chain structure of cellulose makes it able to form fibers similar to other vegetable fibers, such as cotton. The chemical structure of cellulose is given in Fig. 4.23 as $C_6H_{10}O_5$ repeated n times, where n is a very large number between 8000 and 10,000.

The hemicelluloses are chemically similar to cellulose, except that the chain length is shorter, about 150 units of glucose, so that the hemicelluloses are not fibrous, but are gelatins.

The chemical structure of lignin has not been completely determined. Lignin bonds the individual wood fibers together, giving wood its compressive strength.

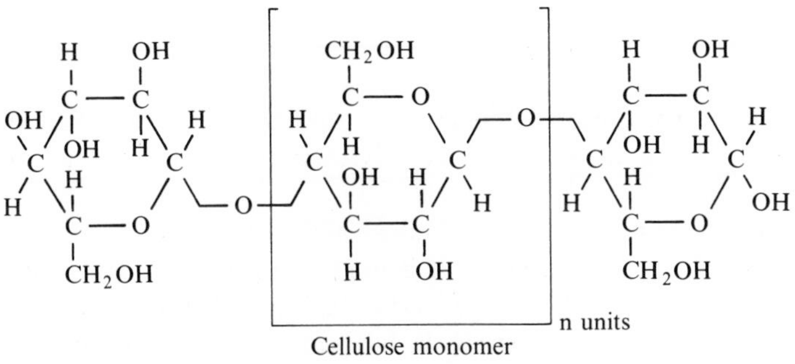

Cellulose monomer

FIGURE 4.23 — Cellulose.

All the properties of wood are influenced by the moisture content of the wood, including hardness and strength. The moisture content of wood in equilibrium with atmospheric moisture is 12 to 15 percent. Variation in moisture content will produce dimensional changes in the wood, though there is little dimension change in the direction of the grain.

The terms *hardwood* and *softwood* do not differentiate between woods that are hard or soft. The hard grain of Douglas fir, a softwood, is extremely hard, whereas balsa wood, a hardwood, is the softest and lightest of all woods. The hardwoods are the deciduous or broad-leaved trees that shed their leaves in the fall. Softwoods are the conifers that bear needles rather than leaves.

Woods are often laminated with plastics, and of course plywoods are bonded with plastic adhesives. In general, plastic adhesives bond better to softwoods and lighter woods such as balsa than to hardwoods and heavy woods. Figure 2.8 shows a balsa-polyester sandwich of great strength and stiffness and very light weight.

A considerable number of plastics are made from cellulose. Usually wood pulp is the raw material. Both thermoplastics and thermosetting plastics can be produced from wood cellulose. The most important of these cellulose plastics are

1.   Cellophane

2.   Cellulose nitrate

3.   Cellulose propionate

4.   Cellulose acetate

5.   Cellulose acetate-butyrate

6. Methyl cellulose

7. Ethyl cellulose

All these materials, except cellophane, are thermoplastic. Note that none of the cellulose plastics are produced by polymerization, but by chemical modification of cellulose.

Cellophane is a regenerated cellulose and is the same polymer as cellulose. It is, therefore, thermosetting. It is used as a packaging film, and in this form is usually coated with other material, such as cellulose nitrate. Cellulose film has excellent clarity and strength, but tears rather easily.

Cellulose nitrate, also known by the trade name Celluloid, is chemically similar to nitrocellulose, an explosive. It has the serious disadvantage that it burns furiously, and is not therefore acceptable for many products, including building materials. It was the first of the synthetic polymers, having been discovered just over 100 years ago. Cellulose nitrate is molded into small articles such as dice and table tennis balls, but is chiefly used in fast-drying lacquers. It is not molded by standard injection-molding and extrusion methods, because the required processing temperatures present a serious fire or explosion hazard.

Cellulose acetate is used as a thermoplastic adhesive, as packaging and photographic film, and for molded articles. It is a clear, tough, and scratch-resistant material.

Butyrate and propionate are also tough thermoplastics. Cellulose-acetate-butyrate is molded into armrests, automobile steering wheels, telephone bases and tackle boxes; cellulose propionate makes excellent safety glasses for face protection.

Methyl and ethyl cellulose are chiefly employed in lacquer and other such formulations rather than molded products.

[4.14.]
special chain structures

Except for the cellulosics of the previous section, all the thermoplastics so far discussed have been based on a carbon chain with various side atoms or groups. Cellulose has a chain composed of carbon rings connected by oxygen atoms (Fig. 4.23). Nylon 6/6 contains a nitrogen atom at every seventh position in the chain, the other six positions being occupied by carbon atoms. In general, a carbon chain made of ethylene groups of two carbon atoms has limited temperature resistance; ethylene groups can be broken out by high temperatures and then can lose hydrogen atoms. The resulting unsaturation leads to oxidation or other types of degradation.

The introduction of oxygen into the chain results in polymers with increased temperature resistance and increased toughness. Two outstanding polymers of this kind have wide application: acetal (polyformaldehyde) and polycarbonate (Fig. 4.24).

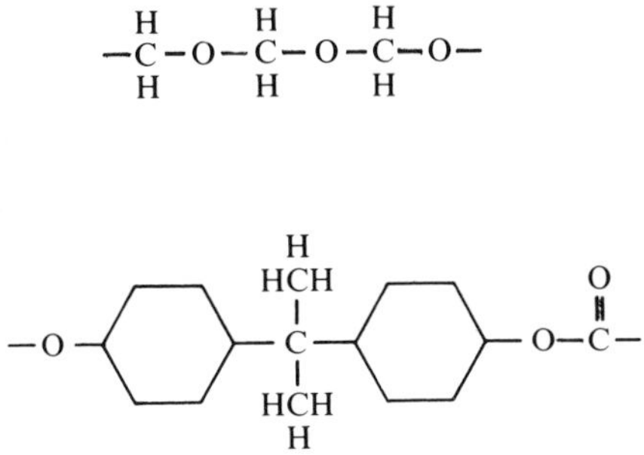

$$\begin{array}{ccccccc} \text{H} & & \text{H} & & \text{H} & \\ | & & | & & | & \\ -\text{C}-\text{O}-&\text{C}-&\text{O}-\text{C}-\text{O}- \\ | & & | & & | & \\ \text{H} & & \text{H} & & \text{H} & \end{array}$$

FIGURE 4.24 — Acetal and the polycarbonate monomer with its two-ring structure.

Acetal (trade name Delrin) has the slippery feel of nylon, which it somewhat resembles. This is a strong and stiff thermoplastic with excellent dimensional stability. Like nylon and polycarbonate, acetal is injection-molded into hardware items such as gears, bearings, cams, hinges, and business machine housings.

Polycarbonate provides a combination of transparency, remarkable toughness, heat resistance, dimensional stability, and flame resistance (unlike acetal). Polycarbonate is the choice for articles that may be subjected to heavy blows, hard hats for example. Except in the presence of stress-raisers such as sharp corners, it is extremely difficult to break polycarbonate. As a glazing material for windows it is scratch-resistant; such a window is virtually impossible to wreck by violent methods. Because of its heat resistance, it must be injection-molded at high temperatures and pressures.

Other linear polymers with oxygen atoms in the chain include polyethylene glycol, chlorinated polyether, polypropylene glycol, polyurethane, and polyethylene terephthalate (Mylar).

Other methods of producing temperature-resistant plastics use sulfur (polysulfones) or silicon (silicones) in the polymer chain. The silicones are partly inorganic (silicon oxide) and partly organic ($-CH_3$), with useful properties at temperatures of about 300° C. They will be discussed under the general topic of elastomers (rubbers), though not all silicone formulations are elastomers.

The polysulfones have the chemical structure of Fig. 4.25. The sulfur atom is in the condition of highest possible oxidation. Therefore, this polymer is highly resistant to oxidation up to 500° C, though severe loss of other properties occurs about 160° C. It is rated self-extinguishing. Polysulfone is not suited to exterior use because of its limited resistance to ultraviolet radiation.

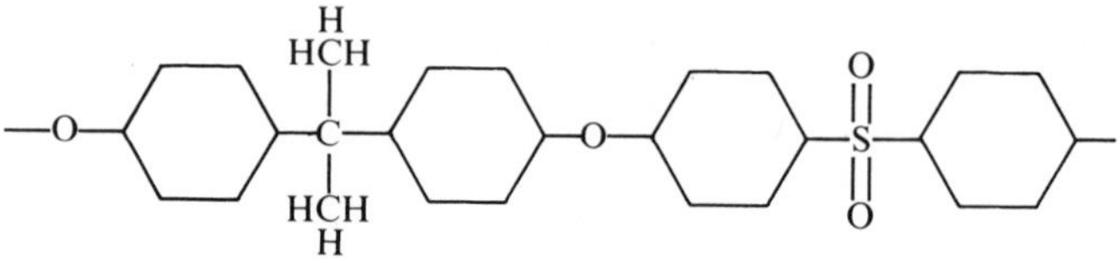

FIGURE 4.25 — Polysulfone monomer.

The sulfone chain contains phenyl (benzene) rings alternating with oxygen atoms and the occasional sulfur atom. The oxygen atom provides increased temperature resistance and toughness. The incorporation of phenyl rings into the chain produces a remarkable increase in temperature resistance, as illustrated in Fig. 4.26. The polystyrene monomer has an aromatic ring attached to one of the carbon atoms. The attached aromatic does not contribute either temperature resistance, strength, or ductility in this location. The same figure shows a unit of poly(p-xylylene), which incorporates the

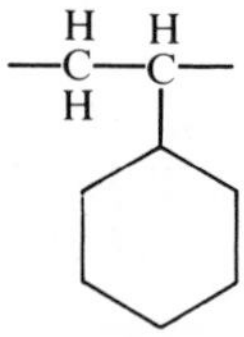

Polystyrene
Melting point 240° C

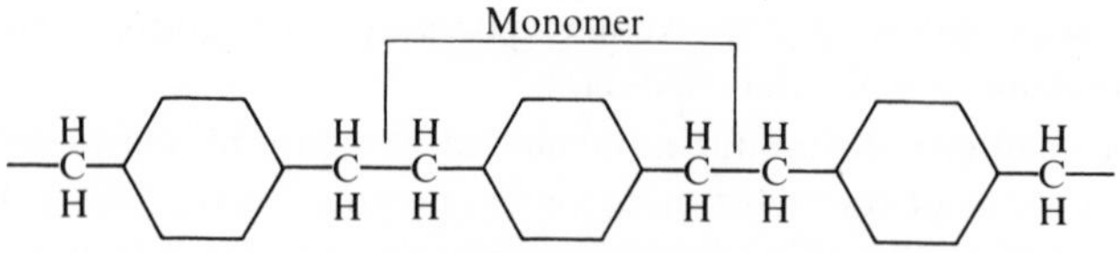

Poly (p-xylylene)
Melting point over 400° C

FIGURE 4.26 — Aromatic ring in branch and in chain.

aromatic ring within the main chain. The improvement of properties is remarkable.

A great many new and research polymers with an aromatic ring in the chain have been developed, including aromatic nylons. Some of these materials, though nominally thermoplastic, are actually infusible (PTFE is also infusible). Theoretically, then, a chain of benzene or similar aromatic rings would be an ideal plastic. This plastic, actually polyphenylene (Fig. 4.27), is, however, difficult to polymerize to a sufficiently high molecular weight. It does not melt even at 530° C, and sustained temperature of 240° C appears not to harm it. The problems of forming such a plastic are severe; clearly,

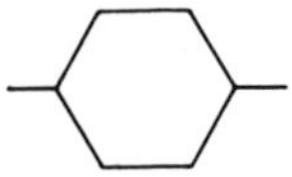

FIGURE 4.27 — Polyphenylene.

such a "thermoplastic" cannot be injection-molded, extruded, or blow-molded. Fluorophenylenes and bromophenylenes have also been researched, but are not commercially available. Polyphenylene oxide and polyphenylene sulfide are, however, in use (Fig. 4.28). In addition to remarkable temperature resistance, these materials are unusually strong and stiff. Since many of these newer thermoplastics are infusible, it is preferable to name them "linear" polymers.

FIGURE 4.28 — Polyphenylene oxide and sulfide.

Comparing these temperature ratings with the metals, we may note here that aluminum has a temperature rating of about 150° C, and carbon steel about 450° C.

Still another method of making superpolymers is to use a ladder structure rather than a linear structure. The ladder type is illustrated for polyacene (not commercially available) in Fig. 4.29. The aromatic rings are bonded to each other in two carbon positions. The breaking of any single bond by heat or other attack does not split the chain, since a single bond still remains. The principle may be extended to make sheet polymers, as illustrated

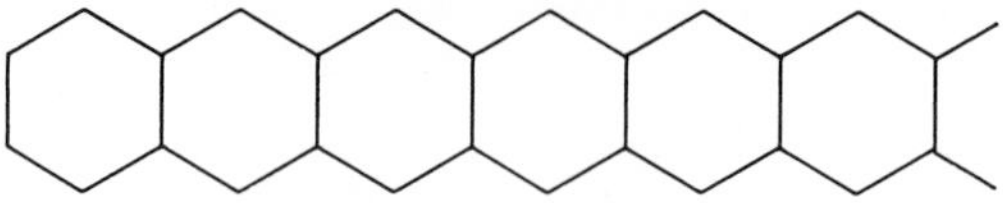

FIGURE 4.29 — Polyacene.

in Fig. 4.30. Such ladder polymers are not yet available in industry, except for partial ladders such as polyimide, discussed below.

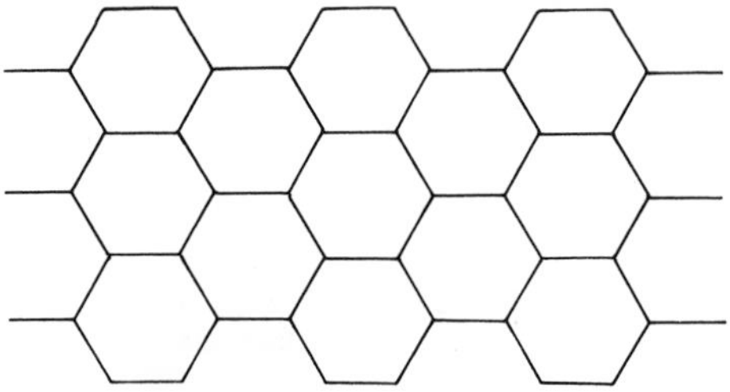

FIGURE 4.30 — A sheet polymer.

Inorganic polymers employing phosphorus, boron, or other elements should in theory provide even higher temperature resistance, but with the exception of silicones such polymers have not shown properties to match those of the linear and ladder types. The ionomers use metal ions to cross-bond between polymer chains.

The ladder polymer that has seen widest use is polyimide (Fig. 4.31). The polyimides are employed as varnishes, adhesives, and molding compounds. The polyimide coating resins require additives to improve ultra-violet resistance. Dupont's polyimide film, named Kapton, has an ultimate tensile strength of 25,000 psi and elongation of 70 percent. It is stable in air to 420° C, is infusible and does not burn.

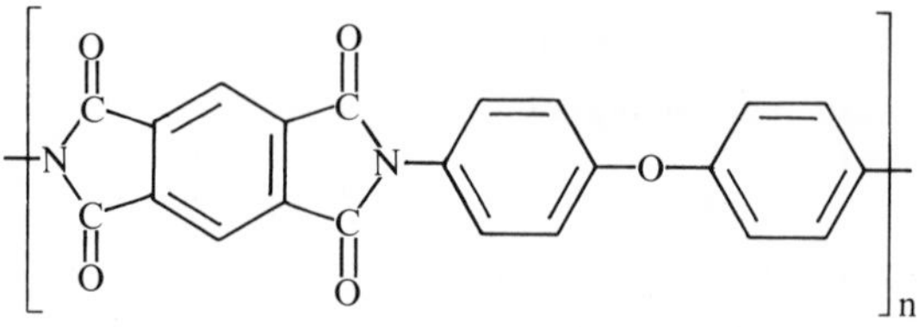

FIGURE 4.31 — A polyimide. Note the ladder type of structure in the left half of the monomer.

With so much chemical versatility to be exploited, it is difficult to pre-dict the capabilities of polymers still to be developed.

[4.15.]
characteristic properties of the thermoplastics

Table 4.1 is a summary of the properties of common thermoplastic materials. Since all these basic materials are available in many formulations, the data supplied are only approximate. Burning rate and impact resistance especially will vary with the formulation. A thermoplastic supplied as film is stronger than the same plastic in a solid section. Nevertheless, the table supplies order-of-magnitude information and a useful comparison between these materials.

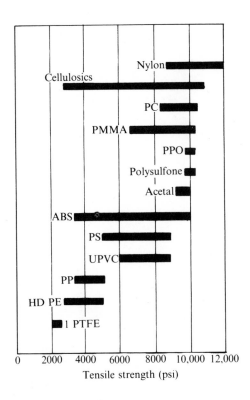

FIGURE 4.32 — Tensile strength of thermoplastics.

Table 4.1
THE COMMON THERMOPLASTICS

	ABS	PE	UPVC	PS	ACRYLIC	POLYCARB	PVF
1. Specific gravity	1.04	0.95	1.4	1.05	1.2	1.2	1.5
2. Tensile strength, psi	5000	$2-4 \times 10^3$	6000	5000	8000	9500	18,000
3. Elongation, percent	40	40–100	10	1.2	5	100	200
4. Modulus of elasticity	250×10^3	$25-100 \times 10^3$	350×10^3	50,000	400×10^3	350×10^3	300×10^3
5. Impact strength	good	good	poor	poor	low	high	high
6. Thermal expansion, per °F	0.00005	0.0001	0.00003	0.00004	0.00005	0.00004	0.00003
8. Resistance to heat, °F	160	200	150	160	150	250	120
8. Burning rate, in./min.	1½	3	none	10	2	1	slow
9. Effect of sunlight	slight	serious	slight	serious	slight	slight	none
10. Type of thermoplastic	flexible	flexible	rigid	rigid	rigid	flexible	flexible

The values for ultimate tensile stress are those obtained in the usual short-time tension test, and give no indication of creep properties. The ultimate strength of most of these materials will decrease slightly over the years of service.

As a rule of thumb, the thermal expansion of thermoplastics and thermosets is about ten times that of steel, and E-values are about one-hundredth that of steel.

The effect of sunlight (ultraviolet) on thermoplastics is difficult to summarize. The table indicates that sunlight has little effect on UPVC. The statement is not true for PVC, which is not stabilized against such degradation, but if stabilized with pigments and ultraviolet absorbers such as titanium dioxide, PVC without plasticizers is one of the most weather-resistant of the thermoplastics. Carbon black is by far the best additive for ultraviolet protection, but cannot be used if the plastic must be pigmented.

QUESTIONS

1. What is a petrochemical? A hydrocarbon?

2. The chemical intermediates obtained from petroleum and natural gas are largely paraffins. Those obtained by distilling coal are not paraffins. What type of intermediate is obtained from coal? Consult a suitable reference.

3. What is meant by polymerization?

4. Sketch the chemical structure of benzene, methane, ethane, ethylene, butane, and pentane.

5. What is the difference in structure between a normal paraffin and an isoparaffin? Illustrate with butane as an example.

6. What is an olefin? An aromatic? A chemical intermediate?

7. Differentiate between a thermoset and a thermoplastic.

8. Why is neither the highest nor the lowest possible molecular weight used in a thermoplastic?

9. Why can a higher molecular weight be used for extrusion than for injection-molding?

10. What is a plasticizer?

11. Why do higher molecular weights give better resistance to ultraviolet in polystyrene and polyethylene?

12. When a thermoplastic cools in a mold, will the shrinkage be greater or less for a crystalline plastic as compared with the same plastic in the amorphous condition?

13. What advantages does crystallinity in a thermoplastic offer?

14. What is meant by an oriented plastic?

15. What thermoplastics are favored for insulating electric wire?

16. Why does high-density PE exhibit greater shrinkage when cooling from the molding operation?

17. What component in wood gives it (a) tensile strength? (b) compressive strength?

18. Is wood a polymer?

19. List some thermoplastics with benzene in the chain.

20. Sketch a ladder polymer.

21. If the glass transition temperature is above room temperature, plasticizers are not useful. Why?

22. What is the difference between an amorphous, an oriented, and a crystalline plastic?

23. Which thermoplastics make suitable glazing materials. Why is polystyrene not suitable for this purpose?

24. Name some thermoplastics resistant to ultraviolet degradation.

25. How do metals and thermoplastics compare in: (a) thermal expansion? (b) modulus of elasticity?

26. What is a fluorocarbon? A chlorinated hydrocarbon?

27. What modifications can be made to the ethylene chemical formula to make it self-extinguishing?

INVESTIGATIONS

1. What synthetic materials are often substituted for paper (including waxed paper) in drafting, wrapping of cigars, etc.?

2. Discuss the advantages of PVC as a rain gutter for houses, as compared with galvanized steel.

3. Attempt to find a stain that cannot be removed from PVF film, if detergent and water are used as a cleaner.

4. Set up a stress-cracking condition (using a suitable liquid solution, for example) in a polystyrene tumbler. A period of time may be required until failure.

5. Clamp a rectangle of polycarbonate in a vise and try to break it, using a hammer or other means. Next make a slight notch in one side of a similar piece of PC and break it in the same manner. Finally, thoroughly wipe the surface of another piece of PC with a strong solvent and break it. Draw your conclusions.

6. PTFE is called a thermoplastic. Try to melt this material.

7. A polyethylene syringe must be sterilizable in boiling water. Would high or low density PE be specified? Is PP suitable for syringes to be sterilized in boiling water? What thermoplastic is used in disposable syringes? Try these materials in boiling water.

8. Certain thermoplastics are used as implants within the human body, in artificial heart valves and artificial hip joints, for example. There is a double problem in these applications. The implant plastic must not release materials toxic to the human body. Also, the human body is highly corrosive to implant materials. Consider which of the following materials you think might serve as implants: (a) polyethylene; (b) UPVC; (c) PS; (d) PTFE; (e) PVF. Would you use a plasticizer in an implant thermoplastic?

9. A sprayable PVC for repair of roofs is offered to you, with tensile strength 750 psi and 250 percent elongation. Is the material plasticized or unplasticized, and would you recommend it as a roofing material?

5

thermosets and elastomers

[5.1.]
thermosetting plastics

Thermosetting plastics, such as wood, wool, Bakelite (phenol-formaldehyde), and epoxy, are not softenable after polymerization. The polymerization process giving a thermoset is not the addition polymerization that produces a thermoplastic, but *condensation polymerization*. In condensation polymerization, a chemical compound reacts with itself or another compound, the reaction releasing or "condensing" some small molecule such as water. Such polymerization may be illustrated by the reaction between phenol and formaldehyde to produce phenol-formaldehyde, shown in Fig. 5.1. Here water is released. A few of the more complex thermoplastics, such as nylon, are

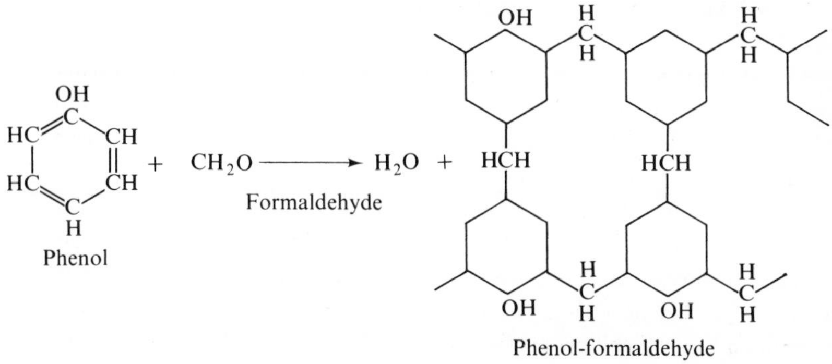

FIGURE 5.1 — Condensation polymerization of phenol-formaldehyde.

also produced by condensation polymerization. Condensation, however, is not a characteristic of the polymerization of the polyesters, familiar in fiberglass-reinforced plastics.

A completely polymerized thermoset could not be shaped except by machining, since all thermosets are brittle. Therefore, these materials must be supplied in some intermediate stage of polymerization. They are classed according to the degree of polymerization:

A stage fusible and soluble, that is, thermoplastic. An
  A stage resin is called a novolac.
B stage fusible and partly soluble
C stage full polymerization

The thermosets are generally brittle, though the elastomers (rubbers) are not brittle even though thermosetting. The freedom of movement of the polymer molecules under stress is greatly limited by cross-linking between the polymer chains. Thus Fig. 5.1 shows two molecules of phenol-formaldehyde cross-linked with carbon atoms. In the case of rubbers this cross-linking is sometimes called vulcanization, the cross-linking agent being usually sulfur for rubbers. Linseed oil and most paints and varnishes harden as oxygen from the atmosphere cross-links the molecules. The effect of cross-linking is to make a three-dimensional network polymer, reducing the plasticity of the material. The number of cross-linking bonds is usually relatively small compared with the total number of cross-linking sites, and the rigidity and hardness of the material is proportional to the amount of cross-linking. The flexibility of a thermoset, therefore, can be controlled either by reducing the number of cross-linking sites (by copolymerizing with a monomer that has no cross-linking sites, for example) or by controlling the number of cross-links.

Most thermosets and rubbers require filler material to improve their properties. Carbon black is commonly added to rubbers and sometimes to thermosetting plastics. Without suitable fillers, thermosets are brittle and lack impact resistance, and when molded have a high shrinkage due to "condensing" of water; this leads to blistering during molding. The epoxies, however, do not give water as a byproduct of curing. Glass fiber, wood flour, calcium carbonate, asbestos, aluminum powder, and many other fillers are employed in thermosetting plastics. In addition to shrinkage control, the physical and electrical properties of the plastics are strongly influenced by the amount and type of filler material. The following is a representative formula for a thermosetting molding compound:

	WEIGHT, LB	WEIGHT, %
Polyester resin	120	30
Styrene (solvent)	2.4	0.6
Calcium carbonate (filler)	190	47.5
Pigment	12	3
Aluminum stearate (mold release)	2.4	0.6
Benzoyl peroxide (catalyst)	1.2	0.3
Fiber glass	72	18
	400	100

Note that in this formulation the weight of filler is double the weight of the resin. The *catalyst* is any compound that will initiate curing of the resin (polymerization). In addition, promoters, inhibitors, and stabilizers may be added to control curing. A promotor or activator will accelerate the catalyst action to produce more rapid polymerization. Inhibitors are used to delay curing so that the material may be stored for a period of time, or to control the rate of curing during molding. Stabilizers delay curing at room temperature but do not prevent curing at molding temperatures.

[5.2.]
phenol-formaldehyde

Since so many polymer materials have been invented, it was inevitable that some of the earlier plastics should decline in importance. PF was the first of the synthetic plastics, being patented in 1909, but it has continued to compete against newer resins. Other formaldehyde plastics are competitors of PF, including urea-formaldehyde and melamine formaldehyde. But phenol-formaldehyde is the cheapest of the thermosets. PF can sustain temperatures of 300° F, or as high as 600° F in the case of heat-resistant grades. Its dimensional stability is outstanding. However, it has the limitation that it can be supplied only in brown and black colors. It is familiarly known by its trade name Bakelite. Some of its many uses include:

1. Wall plates for 110-volt toggle switches and the toggle itself. If these parts are white or pastel colors, then UF is used.

2. Components of fuse boxes, circuit breakers, and electric motor starters.

3. Lamp bases and receptacles.

4. Radio and appliance knobs.

5. Handles for cooking pots.

6. Toilet seats.

This thermoset is used as the adhesive for exterior grade plywoods and is easily recognized in this application by its dark brown color. About four percent of the weight of a sheet of plywood is represented by the PF adhesive. A plywood bonded with PF will not delaminate under conditions of severe exposure to weather and water.

Figure 5.2 shows the stress-strain curve for a typical grade of phenol-formaldehyde; it is elastic to rupture. Such behavior is characteristic of the formaldehyde and epoxy thermosetting plastics. The following are approxi-

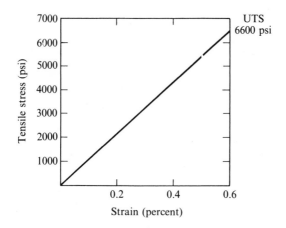

FIGURE 5.2 — Stress-strain diagram for a typical phenol-formaldehyde. This is a brittle and elastic plastic.

mate properties of the several formaldehyde resins. These properties may be altered by fillers and reinforcing materials.

Specific gravity	1.5
Ultimate tensile strength	8000 psi
Elongation	0.5%
Modulus of elasticity	1×10^6
Impact resistance	poor
Thermal expansion	0.0002 in./in. $-°$ F

The formulations of PF in most common use are general-purpose and impact grades. The GP or general-purpose phenolics are suitable for electrical components, knobs, pot handles, etc. This grade incorporates a purified and cooked wood flour as filler material. Impact grades contain fibrous fillers, which intermesh and thus distribute shock loads through the component. The fibers may be glass, cotton, rayon, nylon, or paper pulp. Tool handles and gears are made of impact grades.

[5.3.]
amino thermosetting plastics

Urea-formaldehyde and melamine formaldehyde are both derived from ammonia and are referred to as amino plastics. These thermosets do not have the excellent dimensional stability of PF nor its high heat resistance, but both have the advantage that they may be colored white, pastel shades, or any other color. UF does not provide good scratch resistance, while MF is the hardest of all the plastics, with outstanding scratch resistance. UF, being cheaper than melamine, is substituted for PF where color is a requirement, while melamine is familiar in dinnerware. Arborite, Micarta, and other counter-top laminates are a familiar application of UF and MF.

Urea-formaldehyde replaces PF in plywoods where the dark color of PF would not be acceptable; UF, however, does not have the water resistance of PF, so a plywood bonded with UF is not suited to exterior use.

[5.4.]
epoxies

The epoxy resins have oxygen and aromatic rings in the polymer chain. As explained in the previous chapter, such a structure implies superior properties. Epoxies are available in a wide range of formulations to provide a range of properties and curing characteristics. They are usually supplied as two components to be mixed and cured, but one-component epoxies are available that cure by the absorption of oxygen. No water or volatiles are produced in the curing cycle.

The epoxies are strong materials, with outstanding adhesion to most surfaces and excellent corrosion resistance. Shrinkage is very low during curing, so they are suitable as filler-adhesives. Their range of use is very broad and includes adhesives and surface finishes. Construction uses include the bonding and repair of concrete and concrete floors, traffic-bearing surfaces, and epoxy terrazzo floors. The electrical industry uses epoxy for potting and encapsulating of electrical hardware and for electrical insulation. Aircraft construction requires epoxy adhesive for bulkheads, floors, wing flaps, and various sandwich panels. Tooling epoxies, often including aluminum powder as filler, are cast to make large and complex forming dies such as the door-forming dies of Fig. 5.3. Tooling epoxies are much easier to machine than tool steels and are readily repaired or altered by adding additional epoxy resin.

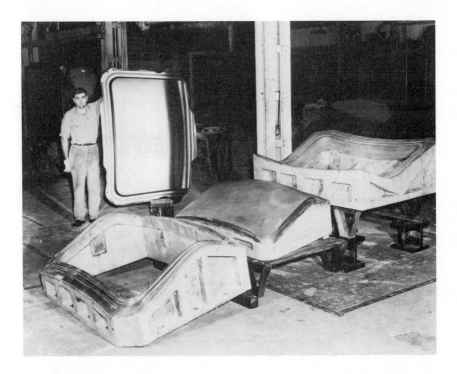

FIGURE 5.3 — An epoxy metal-forming die producing refrigerator doors. *(Courtesy of Rezolin, Inc.)*

[5.5.]
polyesters

Thermoplastic polyester is molded into articles requiring high heat resistance or impact resistance, such as small gears and other machine parts. Mylar film also is a thermoplastic polyester. Polyesters, however, are more familiar as thermosetting plastics for the manufacture of boats, auditorium seats, fishing rods, automobile bodies, and an endless list of other familiar uses. In virtually all these applications the polyesters are reinforced, usually with glass fibers.

Unlike almost all other thermosets, the polyesters polymerize rapidly at room temperature without pressure. They are a large family of condensation polymers made from saturated and unsaturated organic acids and alcohols and cross-linked by styrene, acrylics, or other monomers by means of a suitable catalyst. A large number of acids and alcohols may be selected for the polymerization (actually copolymerization) of polyesters. An alcohol that contains more than one hydroxyl (OH) group is termed a polyhydric alcohol; one that contains two OH groups is termed a dihydric alcohol. Similarly, a dibasic or a polybasic acid contains two or more than two hydrogen atoms that can be replaced by hydroxyl groups. A polyester is the result of a reaction between a dihydric alcohol, and a dibasic acid, for example, maleic acid or fumaric acid and propylene glycol or ethylene glycol. If the reaction is between a polyhydric alcohol and a polybasic acid, the result is an *alkyd*; the alkyds are one of the basic types of paint. An alkyd is, therefore, one type of polyester.

The polyesters are cured rapidly by the intervention of a small quantity of catalyst, usually a peroxide such as methyl ethyl ketone peroxide. Curing is exothermic, that is, producing heat, and the exotherm accelerates the curing action. The acid or alcohol (glycol) has olefinic bonds $(C = C)$, and cross-linking occurs at these unsaturated sites. Frequently the polyester resins are supplied as solutions of unsaturated polymers in vinyl monomers. The vinyl provides crosslinks and also controls the viscosity of the formulation. Inhibitors are added to prevent premature gelling while the resin is in storage. The cure cannot be delayed indefinitely; polyesters must be used within six months or some other recommended period of time. At the expiration of the storage life the inhibitor is used up, and polymerization then progresses more rapidly. The period during which the resin remains ungelled is called the induction period; the length of this period depends on temperature. At an elevated temperature such as 300° F this period may be seconds only (Fig. 5.4).

When the components react, the polymer occupies less volume than its components. This explains the shrinkage that occurs on curing. The exotherm

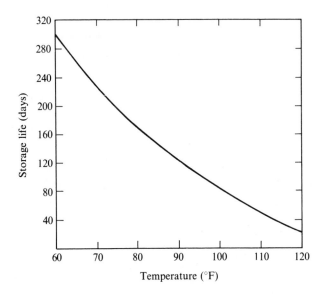

FIGURE 5.4 — Typical curve of storage life vs. temperature for a polyester resin.

produces higher curing temperatures in thick sections than in thin sections, and higher temperatures result in increased shrinkage. Fillers and reinforcements reduce the amount of shrinkage, since they absorb some of the exotherm and do not themselves shrink.

An almost limitless number of polyester formulations is possible, and hence the formulation may be made to suit the requirements of almost any product. Dozens of suitable acids and alcohols are available. There is also a wide selection of suitable cross-linking monomers in addition to styrene, and these may be used singly or in combinations of two or even more. The variety of choices extends even to catalysts and promoters. Even the range of fillers is extensive, and the properties and proportions of any specific filler may be varied as required. For example, glass fibers may be woven or loose, short or long, etc. No other polymer system offers the remarkable versatility of the polyesters.

There are, as always, some disadvantages. The limited shelf life already noted is a limitation that is not shared by some other thermosets such as the epoxies. Some components of the polyester formulation, particularly catalysts and styrene, require careful handling because of fire hazard. Contact of certain polyester catalysts such as methyl ethyl ketone peroxide with certain hardeners and promoters may generate explosions. The generation of static electricity and the risk of dermatitis must be prevented during handling operations.

High vapor concentrations of volatiles such as styrene also are a hazard to health.

Polyester formulations thus are a complex subject to discuss. In summary, however, a general-purpose formulation may be described briefly as an unsaturated polyester dissolved in styrene. The styrene cross-links the polyester, though it is equally logical to say that the polyester cross-links the styrene. The structure of the cured polyester is suggested by Fig. 5.5.

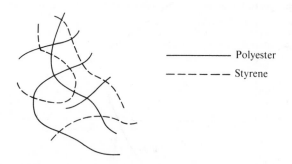

      ——————— Polyester

      — — — — — — Styrene

FIGURE 5.5 — Typical cross-linked structure of a polyester.

Improved weathering characteristics are given by using a 1:1 mixture of styrene and methyl methacrylate.

Table 5.1 compares properties of an unfilled and unreinforced polyester with a mineral-filled polyester also unreinforced. The properties of a reinforced polyester are strongly influenced by the amount of reinforcement, and will be discussed in a later section. The data should be considered as representative only.

Table 5.1

	CAST UNFILLED POLYESTER	MINERAL-FILLED POLYESTER
Tensile strength	8,000 psi	3,500 psi
Compressive strength	20,000 psi	22,000 psi
Flexural strength	13,000 psi	8,500 psi
Elongation	5%	approx zero
Impact strength	very low	very low
Thermal expansion (°C)	0.00007	0.000035
Continuous heat resistance, °F	250	300
Water absorption	0.4	0.45

[5.6.]
general grades of polyester resins

1. *General-purpose.* This type is based on phthalic anhydride, maleic anhydride, and propylene glycol, and is either fast or slow curing. A wide range of physical and electrical properties is possible.

2. *Hand Layup.* Hand layup resins are supplied in low viscosities for good wetting of reinforcing fibers.

3. *Light-stabilized.* These resins are formulated for resistance to weathering, frequently by replacing styrene with acrylics, and therefore are used for exterior panels and similar applications. Additives supply ultraviolet protection and resistance to oxidation and water.

4. *Flame-resistant.* These grades are formulated with chlorine atoms, antimony trioxide, or some other material that will provide self-extinguishment in the presence of a flame. Such grades do not weather well.

5. *Heat-resistant.* This type has high temperature resistance and high heat distortion temperature.

6. *Chemical-resistant.* Usually blended with triallyl cyanurate for broad resistance to chemical attack.

7. *Flexible.* For applications requiring impact resistance.

8. *Resilient.* The resilient formulations are impact resistant but are not flexible. Cross-linking is more limited than is the case for rigid resins, and modulus of elasticity is lower. Heat resistance is usually less also.

9. *Non-air-inhibited.* These resins cure to a tack-free condition on exposure to air and are used for coatings that must resist blushing on exposure to water.

[5.7.]
allyls

In many applications the allylic resins are competitive with the polyesters. They offer certain advantages:

1. Longer shelf life.

2. Less shrinkage during the cure cycle.

3. Somewhat better chemical, optical, and electrical properties.

4. Higher heat resistance.

The most widely used allyls are diallyl phthalate (DAP) and diallyl isophthalate (DIAP). These cure with peroxide catalysts. As with the polyesters, there is a considerable range of choice in monomers for allyl thermosets.

In general, the allyls are more expensive than the polyesters and, therefore, find few uses in consumer goods. They can be molded at lower pressures and in faster molding cycles. Their advantages fit them for industrial products that require their superior physical properties, particularly in electrical applications. Thermal stability of the allyls is superior, ranging as high as 500° F for triallyl cyanurate.

The allylics are also excellent optical materials. Light transmission for these thermosets is closely that for glass or acrylic, and impact and abrasion resistance is superior to both glass and acrylic.

[5.8.]
safety

Safe practices in the handling of materials are developed through knowledge of their characteristics, good judgment, and an awareness of how trouble can develop from simple or trivial circumstances. Inexperience is one of the most common causes of accidents. Safety becomes then an attitude of mind or habit and can hardly be promoted by admonitions in a book such as this. Nevertheless, some remarks must be directed to this subject. Methyl ethyl ketone peroxide will be used as an example, since the safe handling of this material requires both technical knowledge and unusual care. Most plastic chemicals, however, are less hazardous than MEK peroxide.

The suppliers of hazardous materials supply bulletins on the handling of such materials. Immediately upon receipt of such materials these bulletins must be carefully read, understood, and applied. All markings and labels on the container should be noted, and must not be defaced or obscured.

Methyl ethyl ketone peroxide is dangerous by virtue of the fact that it contains the combustible elements carbon and hydrogen, plus also the oxygen required to burn these elements. Since it contains its own oxygen, smothering techniques may not extinguish a fire. This material should be stored in the original container; rehandling may present hazards, and there could be a dangerous contaminant in a different container. Storage temperature should be low, since this material may not be stable at temperatures above 32° C (90° F). Containers must be removed from such sources of heat as steam pipes, radiators, flames, or even radiation from the sun. Isolation from other materials is strongly recommended, especially from strong acids, accelerators, and oxidizable compounds. The storage area must be clean and unheated, and closed to casual and untrained personnel. No smoking is permitted in the storage area, nor is dispensing of the material.

In all storage and dispensing areas a drum of vermiculite or perlite should be available to soak up any spillage. After use such material must be swept, not shovelled, since steel shovels may produce sparks on concrete floors. Consult the supplier for methods of disposing of such contaminated material.

The direct mixing of promoters or accelerators with any peroxide is extremely hazardous.

A face shield should be worn when one is dispensing such peroxides.

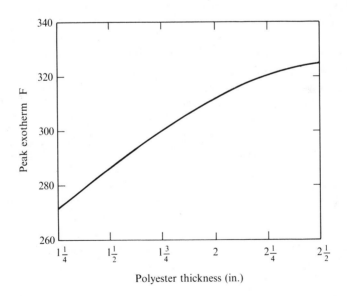

FIGURE 5.6 — Typical peak exotherm temperature for a polyester as a function of casting thickness.

[5.9.]

elastomers

The elastomers or rubbers, like the thermosets, are cross-linked, and, also like the thermosets, are elastic in their behavior. However, the thermosets show little elongation under stress, whereas rubbers are capable of elongations as high as about 1000 percent at tensile failure. The great difference in elongation is explained by the relatively few cross-links in rubbers; thermosets are heavily cross-linked.

The standard stress-strain curve displays the most important characteristic of a material: its stress-strain behavior. A number of such curves have been offered in this book, and they take many shapes. Some kinds of stress-strain behavior are undesirable, such as the poor ductility of most of the thermosets. No matter how strong such materials may be, they would be more acceptable if there were some ductility as an assurance against cracking or unexpected brittle failure. While designers are preoccupied by stress, the necessity for elongation is sometimes overlooked. Many components do not need great strength, but must have flexibility, elongation, and softness. Such characteristics are required in resilient floors, weather-stripping, joint sealants, expansion joints, adhesives, vibration mountings, footwear, vehicle tires, and other applications of similar function. Elastomers are capable of extreme elastic deformation at low levels of stress. The strain is not proportional to stress, as may be seen in the rubber stress-strain diagram of Fig. 5.7. This is a typical S-curve exhibited by rubbers.

In the case of rubbers there is a time lag between stress and strain, sometimes many minutes long. One may observe the lag by indenting a

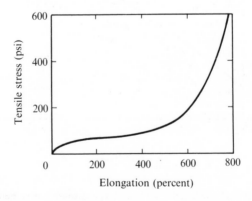

FIGURE 5.7 — Stress-strain curve for a polyisobutylene rubber.

rubber and watching the indention slowly disappear. This characteristic resembles the viscosity of stiff liquids. Unlike the metals and plastics, there is no change in volume when a rubber is strained. Rubbers, therefore, behave remarkably like elastic viscous liquids.

Because they are not cross-linked, the thermoplastics are characterized by plastic strain and creep, whereas the cross-linked thermosets are elastic and brittle. The rubbers are usually addition-polymerized from unsaturated monomers such as butadiene, and cross-linked (vulcanized) with sulfur at approximately every five-hundredth carbon atom. Increased cross-linking gives increased stiffness and reduced elongation.

Some of the more important rubber monomers are shown in Fig. 5.8. With few exceptions, these monomers contain double bonds. These double-bonded monomers when polymerized are susceptible to ozone attack, which results in crazing, and additives must be used to protect the elastomer. Silicone rubber and chlorosulfonated polyethylene are without such double bonds, and as a result weather well. Most rubbers are also susceptible to ultraviolet attack. The usual additive for ultraviolet protection is carbon black, which also increases tensile strength and wear resistance of the rubber.

Like other polymers, the elastomers become brittle at temperatures below

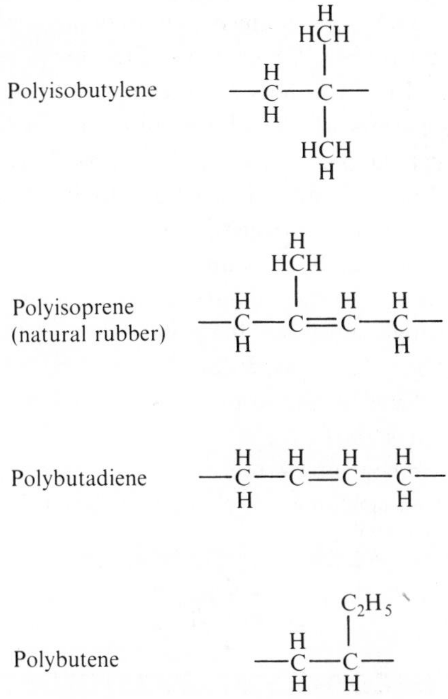

FIGURE 5.8 — Rubber monomers.

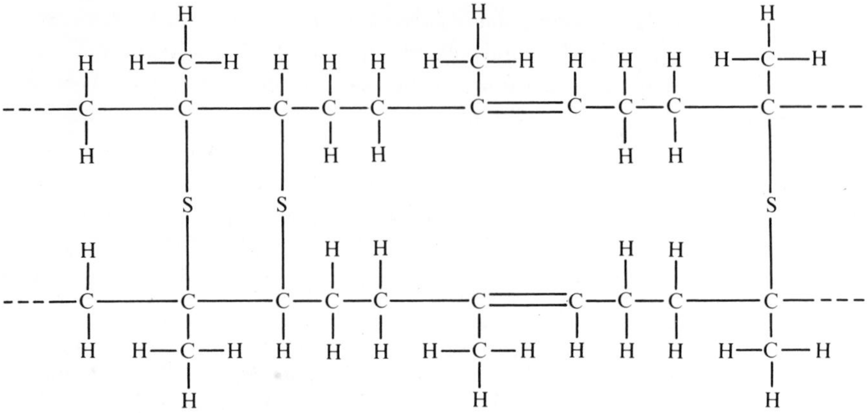

FIGURE 5.9 — Vulcanization (cross-linking) of rubber with sulfur.

their glass transition temperature, which is in the range of $-20°$ C for most rubbers, and about $-60°$ C for natural and silicone rubbers.

Most rubber is molded into articles for transportation. The high friction of rubber surfaces provides traction for soles and heels of footwear, vehicle tires, conveyor belting, and rollers. Rubbers, however, are intimately associated with plastics technology. A foamed polyurethane roof is often protected with a surface of butyl rubber, for example. Elastomers also serve important functions as molds and mold linings for forming plastics.

At least three-quarters of all rubber consumed in this country is synthetic rubber. Natural rubber (polyisoprene) has certain advantages that require its use in a limited range of articles, chiefly vehicle tires, though it has poorer wear resistance than synthetic rubbers for tire use. Although automobile and truck tires are made of synthetic rubbers, chiefly SBR rubber, the larger tires for off-the-road heavy construction vehicles and mining trucks, and aircraft tires (which must not blow out when the aircraft lands) are made of natural rubber. The reason for this is that the heat generated in natural rubber when flexed is not so great as in synthetic rubbers (this heat generation is termed hysteresis).

Only the more important elastomers are discussed in the remarks that follow. Special types of elastomers are not mentioned.

[5.10.]

the common elastomers

1.   *SBR* (styrene-butadiene rubber). This is the general-purpose rubber, used in tires, belts, hose, rubber floor tile, rubber cements, and latex

paints. It is a copolymer of 25 percent styrene, 75 percent butadiene. Tensile strength after vulcanizing and compounding with carbon black is 2500–3500 psi, which is less than the tensile strength of natural rubber, and elongation is 500 to 600 percent. In abrasion resistance and skid resistance it is superior to natural rubber, with better resistance to solvents and weathering.

2. *Natural rubber.* This is polyisoprene, which can also be made synthetically. It is distinguished by low hysteresis and high tensile strength, but is readily attacked by solvents, gasolines, and ozone. Tensile strength is in the range of 3500 to 4500 psi and elongation is 550–650 percent.

3. *Butyl rubber.* The monomer of butyl rubber is isobutylene. The polymer is completely saturated and thus cannot be vulcanized (cross-linked) by standard techniques. Cross-linking sites are provided by copolymerizing with a small amount of butadiene or isoprene. This rubber is impervious to gases, and thus serves as a vapor barrier and hose lining. It is used in the hoses through which polyurethanes are pumped, since if oxygen permeates through the hose it reacts with the resin. Butyl rubber is highly resistant to the agents of outdoor weathering, including oxygen, ozone, and ultraviolet radiation, and is employed as a roofing membrane over polyurethane foam.

4. *Nitrile rubber.* A copolymer of butadiene and acrylonitrile, nitrile rubber has excellent adhesion to metals and resistance to oils and solvents. Its uses include gasoline hose and hose linings, aircraft fuel tanks, and printing rollers, and it is a preferred rubber for O-rings. An interesting use is that of nozzles for aerosol cans to resist the fluorinated hydrocarbon refrigerant gases that pressurize such cans.

5. *Polychloroprene rubber.* This rubber is usually referred to by its trade name, Neoprene. Its resistance to oxidation, aging, and weathering is good, enabling it to be used as a construction material, like butyl rubber. Other characteristics are its resistance to oils and solvents, abrasion, and elevated temperature. Because of its chlorine content, it does not propagate flame. Though best known as an electric cable insulation, polychloroprene has a variety of other uses, including gaskets, hose, engine mounts, sealants, rubber cements, and protective clothing such as gloves and aprons.

6. *Polysulfide rubber* (Thiokol). This is a rubber of low mechanical strength but outstanding resistance to solvents, low gas permeability, and outstanding weathering characteristics. Adhesion to metals is excellent, and this with its other characteristics accounts for its use as a caulking compound and sealant in building construction. It is also used as a roofing membrane. Thiokol and polyurethane rubbers are also used as fuels for solid-fuel rockets.

7. *Acrylic rubbers.* Polyacrylates are cross-linked, but not vulcanized in the usual process by sulfur. Resistance to oils, oxygen, ozone, and ultraviolet radiation is outstanding. These rubbers are used in latex paints and roof membranes.

8. *Rubber hydrochloride.* This material is better known as Pliofilm. It is used in the form of transparent film for the packaging of cheese, meats, and other foods. It is easily identified by its unusual tensile and tear strength.

9. *Reclaim rubber.* Reclaimed and reprocessed rubber is used in rubber cements, weatherstripping, and other general applications.

[5.11.]
silicone rubbers

The silicone monomer is shown in Fig. 5.10. This remarkable monomer has a silicone-oxygen chain with methyl side groups. The silicon-oxygen chain is unaffected by ultraviolet radiation, oxygen, or ozone; however, these rubbers are soft and weak (tensile strength of about 1000 psi), and have limited elongation, 400 percent or less. Tear strength also is poor. Elastomeric properties are retained at temperatures as low as $-200°$ F and as high as $600°$ F. Such polymers have a wide range of uses, including water repellent treatments and mold release agents for plastic, and even high-temperature greases.

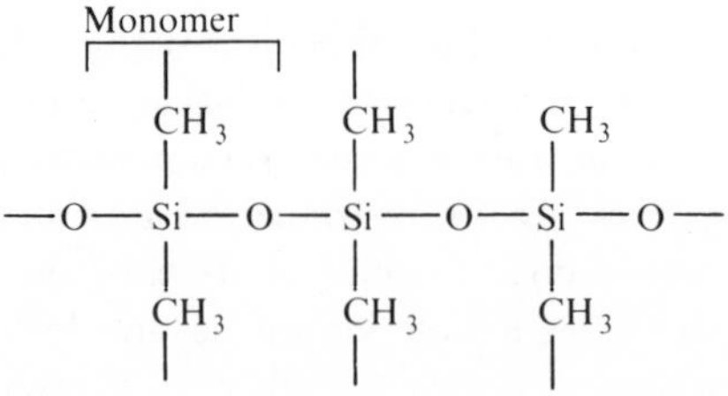

Monomer

$$-O-Si-O-Si-O-Si-O-$$

with CH_3 groups above and below each Si atom.

FIGURE 5.10 — Silicone rubber polymer.

Silicone rubbers can be vulcanized at room temperature. (They are RTV rubbers—room-temperature vulcanizing.) This characteristic, together with very low shrinkage and a temperature limit of $600°$ F make the silicone rubbers a highly useful mold material. They are especially effective as mold linings for poured polyurethane castings, since the two components of the rubber are simply mixed and then applied by spray or paintbrush.

[5.12.]
chlorosulfonated polyethylene

The chemical structure of chlorosulfonated polyethylene, trade name Hypalon, is shown in Fig. 5.11. This is a polyethylene modified by chlorine and sulfonyl chloride. Vulcanization occurs through the sulfonyl chloride groups. The absence of unsaturation indicates excellent weathering resistance, and the chlorine atom contributes flame resistance. Hypalon does not require carbon black as an additive, and thus may be pigmented to give any color, even pastel shades. It is available in sheet form or as a liquid dissolved in solvents. The liquid solution hardens readily to a beautiful gloss; indeed, this must be called a beautiful rubber. (The roof of the terminal at Dulles International Airport near Washington, D.C. is finished in white Hypalon.)

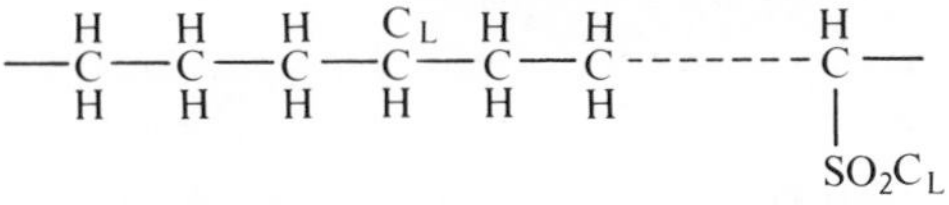

FIGURE 5.11 — Approximate monomer of chlorosulfonated polyethylene. There is a sulfonyl chloride group approximately every 12th carbon atom.

[5.13.]
polyurethane rubber

Polyurethane is available in many formulations and can be given either plastic or elastomer characteristics. Though more familiar as plastic foam or foam rubber, this material is discussed here as a solid rubber.

Polyurethane rubbers, like silicone and Hypalon rubbers, are room-temperature curing. Hypalon is a one-component rubber, whereas the others require two components. The polyurethane rubbers are available in a range of Durometer hardness, generally from about 65 to 100. A typical polyurethane rubber has a tensile strength of 5000 psi, with high elongation, great tear strength, and outstanding abrasion resistance. Tensile strength as high as 8000 psi is possible. The high strength and abrasion resistance of poly-urethane rubber suggest that it would make an excellent vehicle tire material, and so it would, except for its low skid resistance and high hysteresis. Nevertheless, this rubber gives outstanding life when fitted to lift truck and other types of warehouse vehicles as cushion (noninflatable) tires.

Other uses of polyurethane rubber include shoe heels, printing rollers, mallet heads, oil seals, diaphragms, vibration mounts, gears, pump impellers, bowling pins, gaskets, football helmets, and molds for plastics and rubbers. Figure 5.12 shows an imaginative application, the shoeing of a horse with polyurethane "Easyboots." The author has used such shoes on his horses, and doubts that they will ever wear out.

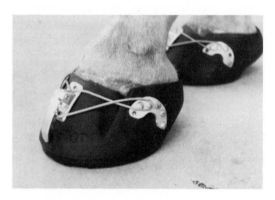

FIGURE 5.12 — Polyurethane "Easyboots."

QUESTIONS

1. What is the difference between a thermoset and a thermoplastic in the following characteristics?
(a) Chemical structure.
(b) Stress-strain properties.
(c) Temperature limits of service.

2. Designate as either thermosetting or thermoplastic: (a) polyethylene; (b) paint; (c) paper; (d) wood; (e) asphalt; (f) PVC; (g) polyurethane; (h) elastomers; (i) shellac; (j) linseed oil; (k) polyvinyl fluoride; (l) epoxy.

3. If a paint or an adhesive hardens by evaporation of a solvent, is the material thermosetting or thermoplastic?

4. What is the usual cross-linking agent for (a) polyesters? (b) varnishes? (c) elastomers?

5. Explain why most paint finishes can be removed by a solvent when wet but not after they have dried (cured).

6. What deficiencies in rubbers are associated with double bonds between carbon atoms in the chain?

7. What rubbers are employed for external exposure in building construction?

8. What influence does the number of cross-links have on the hardness and tensile strength of a thermoset?

9. Suggest a rubber for the following applications:
   (a) A mold lining for polyurethane casting.
   (b) A roofing membrane.
   (c) A high-strength rubber sheet.
   (d) A caulking compound for exposure to weather.
   (e) A rubber cement.
   (f) A road vehicle tire.
   (g) An abrasion-resistant rubber.
   (h) A tank lining subject to low temperatures.
   (i) An impermeable hose lining.
   (j) Electrical insulation.

10. Which is the most scratch-resistant plastic?

11. (a) Why are rubbers abrasion-resistant despite their low hardness?
    (b) Would a hard rubber be more abrasion-resistant than a softer grade?

12. What is meant by cross-linking?

13. Sketch a typical stress-strain curve for an elastomer, a thermoset, and polyethylene.

14. What hazard is involved in mixing a promoter and a catalyst together?

15. What advantages does an epoxy cement offer?

16. What advantages does silicone rubber offer as a mold lining for plastics?

17. For what reasons are fillers and reinforcement added to thermosets?

18. In Sec. 5.1 a formula for a thermosetting molding compound is given. State the use of each of the ingredients.

19. In a polyester formulation, what is the purpose of (a) an inhibitor? (b) a promoter? (c) a catalyst?

20. What advantages does PF offer over other formaldehyde thermosetting resins?

21. How is an impact grade of PF formulated?

22. In what properties does MF exceed UF?

23. What is the meaning of exotherm?

24. For what purpose is methyl methacrylate used in polyester formulations?

25. (a) What is the approximate shelf life of polyesters? (b) What do you suppose the shelf life of epoxy cements to be?

26. In what characteristics are the allyls superior to the polyesters?

27. Which rubber is the general-purpose rubber?

28. (a) In what characteristics is SBR superior to natural rubber? (b) In what characteristics is natural rubber superior to SBR?

29. What is meant by an RTV rubber?

6

additives, fillers, and reinforcements

It is not usual to use a pure resin in a plastic product. The properties and performance of a plastic or an elastomer are usually improved by various types of additives. Even the appearance of a plastic product may be improved by additives; for example, voids within a transparent plastic are unsightly, and these may be hidden if a pigment is added. Many thermoplastics, especially PVC, must be protected by additives from degradation due to the heat of injection-molding and extrusion. Many thermoplastics require antioxidants and ultraviolet absorbers. Most thermosets require either fillers or reinforcing fibers or both.

[6.1.]
solvents

Solvents have many uses in the processing and use of plastics. Plastics must be given a solvent wipe before painting; many plastic adhesives embody solvents. Solvents are used in the maintenance of plastics equipment, such as cellosolve to clean the spray guns that apply polyurethane foams. Plasticizers are blended into certain plastics, and these plasticizers are actually a special type of solvent.

A solvent is best defined as a material used to produce a liquid state in a solid material. The best solvents are small molecules, but since small molecules are volatile, that is, have low boiling points, they have the disadvantage that they evaporate quickly, and if flammable, present a fire and explosion hazard. The chlorinated solvents were developed to avoid the hazards of flammability, but present the additional problem of toxicity. The small molecules of the useful solvents can penetrate easily the large molecules of polymers and, by separating the polymer molecules, dissolve them. Of the many types of solvents, perhaps the following are the more important.

Aliphatic or paraffinic solvents are straight-chain compounds such as heptane and hexane. These have only limited solvency. The *alcohols* also have limited solvency. Typical alcohols are methyl, ethyl, isopropyl, and butyl. The alcohols are paraffins with an attached OH, as shown in Fig. 6.1.

Aromatic solvents contain the benzene ring, such as benzene, toluene, xylene, naphthalene, and styrene. Monomers of thermoplastics can be used to dissolve their polymers. Styrene monomer, for example, is used to solvent-bond polystyrene.

The *ketones* are excellent solvents. They are produced from a hydrocarbon by substituting an oxygen atom for two hydrogen atoms (not terminal hydrogen atoms). The structure of acetone as compared with propane is shown in Fig. 6.1. The usual ketones are acetone, methyl ethyl ketone (MEK), and methyl isobutyl ketone (MIBK). The ketones will dissolve PS, ABS, PMMA, and cellulosics.

The *chlorinated* solvents, such as methyl chloride or carbon tetrachloride, substitute chlorine for hydrogen atoms.

Table 6.1 compares the properties of the paraffinic, aromatic, and chlorinated solvents.

Toxicity — Many of the common solvents are flammable or in the vapor phase even explosive. The chlorinated solvents are not, but are more toxic. Benzene, toluene, xylene, styrene, and other aromatic solvents will dissolve the fatty compounds of the human skin. This action can sometimes lead to

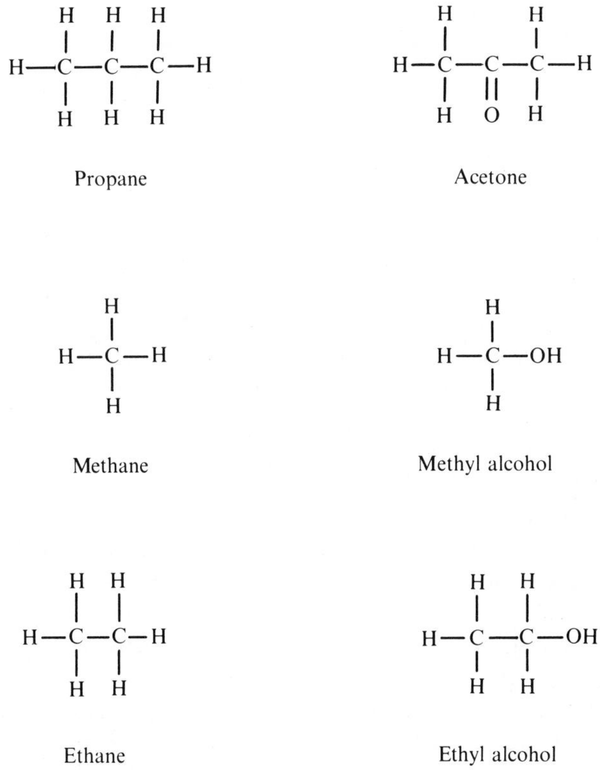

H   H   H
| | |
H—C—C—C—H
| | |
H   H   H

Propane

H   H
| |
H—C—C—C—H
| || |
H   O   H

Acetone

H
|
H—C—H
|
H

Methane

H
|
H—C—OH
|
H

Methyl alcohol

H   H
| |
H—C—C—H
| |
H   H

Ethane

H   H
| |
H—C—C—OH
| |
H   H

Ethyl alcohol

FIGURE 6.1 — Comparison of solvents with their related paraffins.

Table 6.1
SOLVENT COMPARISON TABLE

	PARAFFINS	AROMATICS	CHLORINATED
Cost	low	medium	highest
Solvency	limited	good	best
Flammability	high	very high	none
Toxicity	low	high	high
Paint removal	poor	better	excellent

infection or dermatitis. Carbon tetrachloride is a significant hazard to health, either to lungs or skin.

Acetone is not especially dangerous to health. Methyl alcohol has caused death or permanent damage to the eyes; ethyl alcohol has more transient effects, because the human body readily digests it. Carbon disulfide is a severe nerve poison and a cause of dermatitis.

[6.2.]
plasticizers

A plasticizer is a special kind of solvent. The plasticizers are low-volatility solvents for plastics, often dioctyl phthalate, tricresyl phosphate, or sebacates, which plasticize or soften the plastic. Most plasticizers are incorporated into PVC, dioctyl phthalate being the commonly used one. PVC is a stiff and brittle polymer when unplasticized. The unplasticized PVC (UPVC) is unsuited to such articles as raincoats, garden hose, and gloves, and a plasticized PVC is used in such articles. The plasticizer acts as a lubricant between adjacent polymer chains and thus allows these long molecules greater movement.

Plasticizers are used in thermoplastics either to modify properties such as hardness, stiffness, and ductility (plasticity) or to make the polymer easier to process. A plasticizer may, for example, lower the processing temperature below the temperature at which the polymer will degrade. Plasticizers are not used in polymers with a stiff chain, such as those containing cyclic groups (polycarbonate, for example). Polystyrene is rigid because of its bulky phenyl side group, and for it also plasticizers give no useful effects. The crystalline polymers such as polyethylene and polypropylene rarely use plasticizers.

Plasticizers present the problem of permanence. With prolonged aging, especially at elevated temperatures, plasticizers behave like solvents and evaporate slowly. Therefore, the suitability of plasticized polymers for construction materials is generally doubtful, since such materials are expected to withstand exposure for years without significant change of properties.

[6.3.]
stabilizers

Stabilizers include antioxidants, antiozonants (for protection against ozone in the atmosphere), and ultraviolet absorbers. Degradation of plastics occurs when polymers are exposed to heat, sunlight, and weathering. Such degradation is usually disclosed by a change of color as well as loss of mechanical properties, cracking, and crazing. These property changes occur as a result of chain scission or cross-linking; usually both degrading mechanisms occur.

Polymers made from olefin monomers are especially susceptible to these types of degradation.

Degradation by heat may occur during converting of the thermoplastic, and stabilizers may be required for protection during these heating cycles.

The ultraviolet radiation from the sun is a shorter wavelength than visible light, approximately of the range 3000 to 4000×10^{-8} cm. The energy of this wavelength is sufficient to break most chemical bonds in the polymers. Ultraviolet absorbers act as a screen to absorb such wavelengths. The ideal UV absorber would pass all longer wavelengths and absorb only the shorter UV wavelengths, as in Fig. 6.2. Carbon black is used in polyethylene film and pipe as a UV absorber; of course, it also absorbs visible light. Benzo-

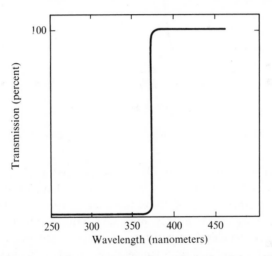

FIGURE 6.2 — Light transmission of an ideal ultraviolet absorber.

phenones and benzotriazoles are also used in thermoplastics. Transmission curves for these stabilizers are shown in Fig. 6.3; these approach the ideal absorption curve.

For heat stabilization, particularly of PVC, dibutyl tin dialurate, lead compounds, and various calcium-barium soaps are employed.

[6.4.]

colors

Most plastics, especially those employed in consumer and commercial articles, are colored. The pigments used in plastics fall into three broad groups: dyes, organic, and inorganic pigments.

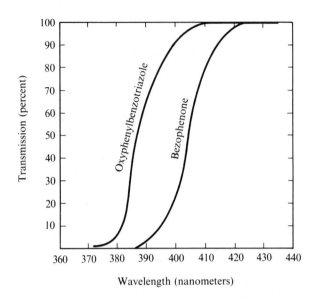

FIGURE 6.3 − Absorption curves for ultraviolet absorbers.

Dyes are aromatic organic chemicals soluble in many solvents, and absorb light selectively so that they produce their characteristic color. They give colored but transparent effects in the clear plastics such as acrylics, cellulosics, and polystyrene. Generally, dyes have limited stability when exposed to sunlight. Their temperature resistance is limited and not necessarily sufficiently high for the molding temperatures used with high-temperature plastics.

Organic pigments are usually insoluble in solvents. They produce opaque effects in plastics. Light and heat stability is better than is the case for dyes.

Inorganic pigments are metal oxides or salts. These have the best hiding power and superior heat and light stability. Titanium oxide is a superior white pigment used in large tonnages in plastics and paints. Iron oxides give yellows, reds, and tans. Cadmium pigments give bright yellows and reds. Carbon black serves as a black color and a UV absorber.

The use of colors in plastics is a complex subject, involving such problems as uniform dispersion of colorant, selecting colors that can withstand processing temperatures without degradation or change of color, and changing colors in molding machines without contaminating one color with another. Selecting pigments for a polycarbonate that requires a molding temperature of 550° F is a more critical problem than a pigment for a room-temperature curing thermoset. Colorants may also influence catalyst action in the use of polyesters and allylics.

[6.5.]
antistatic agents

An electrostatic charge is produced on the surface of a plastic when it is separated from another surface. The other surface may be another plastic part, a mold, or a roller. The static charge thus generated increases as the plastic is handled and transported. These static charges attract dust particles. They may also be a hazard in such areas as hospital operating rooms.

Antistatic agents increase the surface conductivity so that the static charge can dissipate by conduction. External antistatic agents are applied to the plastic surface. They provide only temporary protection, since they can be removed by wear and exposure. Internal antistatic agents are incorporated into the resin compound before molding and function by migrating to the surface at a slow rate. Most of these agents, internal or external, increase surface conductivity by adsorbing water vapor from the air. The effectiveness of the agent, therefore, depends on the relative humidity, since at low humidities less water vapor is available to be adsorbed.

A simple test method for the electrostatic properties of plastics is the ash-tray test. A plastic material is rubbed with a wool cloth and brought close to a layer of cigarette ash. The attraction of ash by the plastic is noted. Because of variants such as relative humidity, this test is often very indeterminate in its results, but it is an interesting one to try.

[6.6.]
fillers

Fillers are a requirement for most thermosets, but are sometimes also used in thermoplastics. Except for vinyl floor tile, a maximum of about 25 percent filler is used in thermoplastics. Fillers may be added to reduce cost by reducing the amount of expensive resin, or to improve certain properties of the plastic product, such as compressive strength or impact resistance. They are often essential in thermosets for reducing mold shrinkage.

The selection of a filler is determined by the physical characteristics inherent in the filler and imparted to the resin and by cost. The filler, however, must have low moisture absorption, must not adversely affect surface finish, and must not cause abrasive wear in processing equipment. The filler also must be capable of being wetted by the resin. Disadvantageous impurities must be absent from the filler, and it must not influence the final color of the plastic product, unless the filler itself serves as a colorant. The effect of fillers on the stress-strain curve of a phenol-formaldehyde is illustrated in Fig. 6.4.

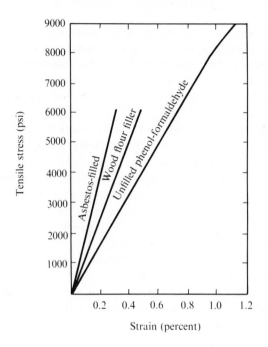

FIGURE 6.4 — Typical stress-strain curves for phenol-formaldehyde formulations.

The following list includes the commonly used fillers. Reinforcing fibers are not included but are discussed in a following section.

Inorganic		*Powdered metal*
talc	clays	aluminum
mica	calcium carbonate	lead
silica sand	titanium oxide	
ground granite	zinc oxide	*Organic*
diatomaceous earth	antimony trioxide	wood flour
aluminum oxide	carbon black	alpha cellulose
calcium silicate		walnut shell flour

Mica improves the electrical properties of plastics; *carbon black* is used as an antistatic or a conductive filler as well as its other uses previously noted.

Talc may be added to thermoplastics to reduce creep. This additive is soft and does not abrade molding equipment.

Zinc oxide protects thermoplastics against weathering; it is also added to polyesters and rubbers. *Antimony trioxide* functions as a flame-resistant additive and also as a white pigment.

Aluminum powder is chiefly used as a filler in epoxy tooling resins for the casting of tooling for metal-forming operations.

Clays (usually kaolin or china clay), *silica sand,* and *calcium carbonate* reduce exotherm in polyester formulations and are useful for maintaining a uniform dispersion of glass fibers in the reinforced resin. Calcium carbonate is a common filler for thermosets and PVC products such as floor tile. PVC floor tile may have calcium carbonate loadings as high as 400 percent. Clays, especially kaolin, are an important ingredient in polyester compounds, giving protection against crazing and some improvement in surface finish. Some PVC formulations also use kaolin. Silica sand is used in construction materials for such products as synthetic polyester brick, stone, and marble.

Wood flour reduces mold shrinkage of thermosets and improves their impact performance. *Walnut flour* results in poorer mechanical properties than does wood flour but gives a smooth and lustrous finsh.

None of these fillers are ground finer than 200 mesh. The surface area of the filler increases rapidly with fineness of the filler, and too fine a powder requires excessive resin to wet the total surface area.

The effect of fillers on the flexural modulus of a flexible polyester is shown in Fig. 6.5.

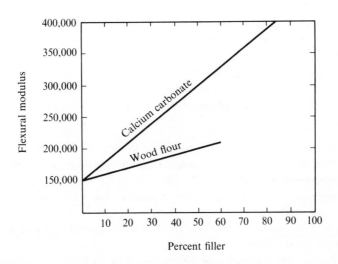

FIGURE 6.5 — Influence of filler on the flexural modulus of an isophthalic polyester.

Filler materials must be clean and free of dust, dirt, iron, copper, and sulfur. In particular, the moisture content of all fillers and reinforcements must be carefully controlled, since moisture is the cause of blistering and other defects. In general, moisture must not exceed 0.5 percent of the weight of the resin. Consider the case of a formulation of 10 percent resin, 90 percent filler, the filler containing 0.5 percent moisture. This amount of moisture is 4.5 percent of the weight of the resin, and is unacceptably large. Clays present the worst difficulties for removal of moisture.

[6.7.]
reinforcing fibers

Fiberglass-reinforced polyester is well-known. Reinforcing fibers of glass and other materials are used also in other thermosets and increasingly in thermoplastics. Glass fiber is the most widely used reinforcement and probably has been used in some application in every thermoset and thermoplastic. The fibrous reinforcements that are most used are these:

Asbestos	Fibrous glass
Cellulose	chopped strand
alpha cellulose	filaments
cotton flock	ribbon
rayon	yarn
Synthetic fiber	mat
polyacrylonitrile	
nylon	
polyester	
polyvinyl alcohol	
popypropylene	

Asbestos — There are six asbestos minerals, all of them fibrous silicates. Chrysotile is the most useful as a reinforcing fiber, and almost all the asbestos used in plastics is chrysotile.

Longer asbestos fibers are more expensive. Short fiber is used for heat resistance as well as reinforcement, whereas long fiber is used for the same purpose in fiber, yarn, or textiles. Little improvement in impact resistance is given by short fibers.

Asbestos felts and mats are distinguished by the thickness of the sheet; 10 mil thickness is a felt, while a greater thickness is a mat. Both are supplied with or without a resin binder, the resin being compatible with the resin to be used in the reinforced product and approximately 5 percent by weight.

The synthetic reinforcing fibers are more expensive than glass or asbestos fibers. Because such fibers are oriented by the fiber-drawing operation, their tensile strengths are very much higher than would be the case for the same polymer in bulk form. Polyvinyl alcohol fibers, for example, have a tensile strength of about 140,000 psi. These and other synthetic fibers give superior impact resistance and excellent adhesion to plastics. The Japanese National Railroad has used polyester railroad ties reinforced with polyvinyl alcohol fiber.

The influence of polypropylene fibers in polyester is illustrated by the graph of Fig. 6.6.

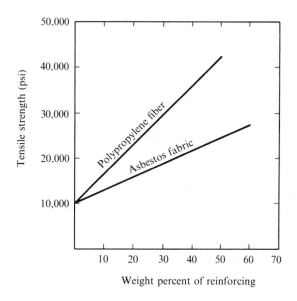

FIGURE 6.6 — Influence of fiber content on tensile strength of a rigid orthophalic polyester.

Wood flour was first used in phenolics and linoleum. Longer alpha cellulose fibers are used as reinforcing, especially for ureas and melamines which must be pigmented.

Cotton flock and cotton linters give somewhat better impact and moisture resistance in thermosets. Cotton cloth (macerated fabric) provides even better impact resistance.

Cotton fabric grades are usually either grade C or L. Grade C is a woven cloth weighing over 4 oz/sq yd with no more than 72 yarns per inch in the fill direction or more than 140 yarns for the total in both warp and fill direction. Heavy cotton ducks up to 20-oz weight are used in the lamination of gears and bearings. Increased fabric weight gives improved impact properties. Grade L is a woven cloth weighing less than 4 oz/sq yd with more than 72 yarns per inch in the fill direction and more than 140 total for both warp and fill. Grade L, or linen grade, is used in such applications as pulleys, small gears, and punching and machining stock. The lighter weight of L cloth results in smoother edges after either punching or machining.

[6.8.]

glass fiber

The total annual consumption of glass fiber-reinforced plastics in this country is of the order of a billion pounds. Such reinforced plastics offer the outstanding advantages of high strength with light weight, and remarkable design flexibility in a range of products from fishing rods to small naval vessels.

Glass filaments do not have quite the same combination of oxides as window glass. Most fiber has the combination

silicon dioxide	52–56%
calcium oxide	16–25%
aluminum oxide	12–16%
boron oxide	8–13%
sodium and potassium oxides	less than 1%
magnesium oxide	0–6%

This is the composition of E glass fiber, the usual grade in reinforced plastics. The E grade (electrical) has good electrical properties, and resistance to attack by water and alkalis, and high strength. C glass (chemical) is low in calcium oxide and alumina, with good resistance to acid attack. It is used in fiberglass surfacing mats. S glass is a high-strength glass fiber, produced in continuous filaments for such applications as pressure vessels. M glass is a high modulus glass; because of its high cost, it has a restricted range of applications.

The properties of fibrous glass are not entirely those of bulk glass, though the modulus of elasticity is the same for both forms of glass. Like any fiber, glass fibers are considerably stronger in a tensile test than bulk glass; it is possible for glass fibers to have tensile strengths of over half a million

CHARACTERISTICS OF E GLASS

Ultimate tensile strength at 50% relative humidity	200,000–220,000 psi
Modulus of elasticity, tension	10.5×10^6
Creep at room temperature	none
Elongation	3–4%

psi. Because of the much higher surface area of fibers, chemical resistance is reduced. Fiber diameters range from about 0.0001 to 0.001 in.

Glass fibers for reinforcing are drawn through a platinum alloy die containing 204 holes. The 204 filaments are gathered into one *end* or *basic strand*. These strands are gathered into rovings of 8 to 408 ends or twisted and plied into yarn.

The glass filaments are fragile and tend to abrade each other when woven. A protective sizing is applied to protect the fibers. This sizing reduces adhesion between glass and resin and must, therefore, be removed and replaced by a glass-to-resin adhesion promoter. The usual coupling agents for this purpose are silicon-organic or chromium-organic compounds. How the coupling agent functions is not entirely understood, but it can be assumed that the organic portion of the adhesion promoter bonds chemically to the organic polymer while the silicon bonds to the silica of the glass; the chromium promoter functions in a similar fashion.

[6.9.]
the influence of reinforcing fibers on strength

Figure 6.7 shows the influence of glass fiber content on tensile strength and flexual stiffness of a polyester. In common with the other graphs of this chapter relating filler content to mechanical properties, the relationship is linear.

Strength of the reinforced part is also influenced by the orientation of the glass fibers. Three cases are possible for orientation:

1.   All glass strands may be laid parallel to each other in a unidirectional reinforcement. This condition can be produced by parallel yarns.

2.   Half the strands may be laid at right angles to the other half, a bi-directional condition that can be produced by woven glass cloth.

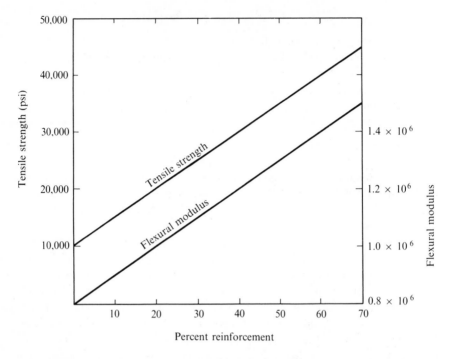

FIGURE 6.7 — Influence of reinforcement on tensile strength and flexural stiffness of a rigid isophthalic polyester, using glass cloth style 181.

3. The strands can be arranged in a totally random manner to make a reinforcement that is isotropic (equally reinforced in all directions).

The unidirectional reinforcement is required in a fishing rod, the bidirectional reinforcement for boats and swimming pools, and the isotropic for safety helmets and office machine housings. In every case the orientation of the fibers is in the direction or directions of highest stress.

The amount of reinforcement that can be used is related to the orientation of the reinforcement. The maximum number of sardines can be packed in a can if they are laid parallel. This principle applies to reinforcing strands also. Glass loadings of as much as 90 percent are possible when continuous parallel strands are used. With bidirectional reinforcement as much as 75 percent reinforcing can be used. A random or isotropic arrangement of fibers allows loadings up to a maximum of 50 percent.

The less expensive types of reinforcement, which are chopped strand, chopped roving, and mat, are used in parts requiring lower glass content and lower strength.

[6.10.]
yarn designations

Because of the very large number of methods by which filaments can be combined for yarn construction, a standard code is used to identify glass textiles. The yarns are made either of continuous fiber, C, or staple, S (short lengths). Staple yarn does not provide high strength. This code is given in Fig. 6.8, and may be explained by the example following Fig. 6.8.

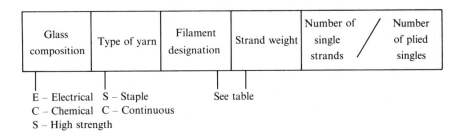

Glass composition	Type of yarn	Filament designation	Strand weight	Number of single strands	/	Number of plied singles

E – Electrical S – Staple   See table
C – Chemical   C – Continuous
S – High strength

FILAMENT DESIGNATION	AVERAGE DIAMETER	STRAND WEIGHT ($\times 100 = $ YD/LB)	FILAMENTS PER STRAND
D	0.00021	1800	51
D	0.00021	900	102
D	0.00021	450	204
B	0.00012	450	408
E	0.00029	225	204
G	0.00035	150	204
DE	0.00025	150	408
B	0.00012	150	1224
K	0.00051	75	204
G	0.00035	75	408
DE	0.00025	75	816
K	0.00051	37	408
G	0.00035	37	816
K	0.00051	18	408

FIGURE 6.8 — Yarn designation chart.

ECD 1800 1/0

E = electrical grade of glass
C = continuous fiber

D = filament diameter 0.00021

1800 = 1800 × 100 = 180,000 yards per pound of the basic strand. From the table, there are 51 monofilaments per strand.

1/0 = a single yarn. The first digit shows the number of original singles twisted and is separated by a diagonal line from the second digit showing the number of these units plied. The designation ¾ would indicate three single yarns twisted together and four of these groups plied together.

Fiber and yarn are available in many forms for reinforcement. In addition, there is a great variety of glass textile weaves. The following are the forms of fiber and yarn.

1.   Hammer milling reduces glass fiber to lengths of ⅓₂ to ⅛ in. for use as a filler rather than reinforcement. Such material goes into cast and molded products in amounts up to 70 percent. This material is called *milled fiber*.

2.   *Chopped strand* is cut to lengths of ¼, ½, 1, or 2 in. for reinforcing molding and premix compounds and for thermoplastic injection molding.

3.   *Continuous strand* is supplied either as yarn or roving. *Yarn* is twisted single-end strands for unidirectional reinforcement, as required in fishing rods, for example. *Continuous roving* is untwisted multistrands. *Spun roving* is a ropelike bundle of strands of continuous glass fiber, used in filament winding of pressure vessels, pipe, rod stock, and similar products. The softness of roving may be adjusted by the amount of size applied and the type of size.

[6.11.]
mats and fabrics

Reinforcing mats are made of either chopped strands or continuous swirl strands laid in nonwoven random pattern. The strands are held together by adhesive resin binders. Mats do not provide as high strength as woven fabric. Unlike weaves, mats are generally used only with polyesters. Surfacing mats are discussed in a following section.

Continuous or spun *woven rovings*, also called roving cloth, are coarse, heavy, and drapeable, heavier than fabrics. They contribute high strength

FIGURE 6.9 — Reinforcing mat.

but are lower in cost than conventional fabrics. Weights range from 15 to 27 ounces per square yard and thickness from 0.035 to 0.48 in. Woven roving is principally used in large structures such as swimming pools and boats.

Woven fabrics are woven from yarns of various twist and ply construction. A wide range of weaves and weights are available. Weights range from 2½ to 40 ounces per square yard, and thickness from 0.003 to 0.45 in.

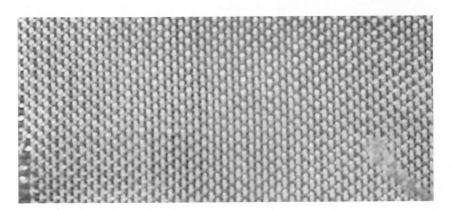

FIGURE 6.10 — Woven fabric.

These glass cloths are available in four standard widths: 38, 44, 50, and 60 inches. In addition, narrow fabric or tape is supplied in widths from ½ in. to 12 in. The threads in the fabric that run parallel to the length of the cloth are called *warp* yarns and are referred to individually as *ends*. The threads running across the width are called *filling yarns* and are referred to individually as *picks*. Thread count is expressed as the number of ends and picks per inch; for example, a count of 57 × 54 means 57 warp threads per inch and 54 filling threads per inch. The first number of the two is always the number of warp ends.

[6.12.]
fabric weaves

The usual weaves for reinforcing fabrics are the following.

1.   *Plain* or *taffeta* (Fig. 6.11). The warp and fill yarns cross alternately. This weave provides fabric stability and firmness with least yarn slip-

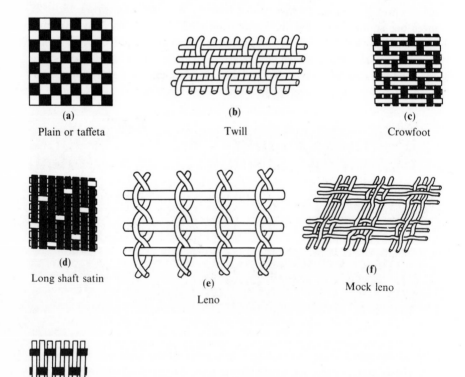

(a)
Plain or taffeta

(b)
Twill

(c)
Crowfoot

(d)
Long shaft satin

(e)
Leno

(f)
Mock leno

(g)
Unidirectional

FIGURE 6.11 — Fiberglass fabric weaves.

page. Strength is provided equally in two directions (if yarn size and count are equal in both directions), and resin penetration into the weave is adequate.

2.   *Basket.* This is similar to plain weave except that two or more warp yarns are woven as one over and under two or more fill yarns. This weave has somewhat better drape and pliability than plain weave.

3.   *Twill.* Figure 6.11 shows a 3 × 1 twill, but others, such as 2 × 1, 2 × 2, etc., are possible. This has better drape than plain or basket weaves but is more difficult to wet.

4.   *Crowfoot.* One warp yarn is carried over three then under one fill yarn. Higher strength is provided in one direction. This weave is used for complex contours and for products requiring higher strength in one direction, such as diving boards and skis.

5.   *Long-shaft Satin.* This weave gives good drape and stretch in all directions but is less open than other weaves. It is used for contoured surfaces such as radar radomes.

6.   *Leno.* The warp yarns are twisted around each other, locking the filler yarns in place.

7.   *Mock Leno.* Yarns run in groups in both warp and fill, locking each other in place where they interlace. This pattern gives good flexibility and drape, and the open pattern promotes good penetration and bonding of the resin.

8.   *Unidirectional.* This is a plain, basket, or satin weave with a higher number of strong yarns in one direction, usually the warp. This weave is used in products requiring strength in one direction, such as diving boards.

[6.13.]
surfacing mat

Any weathering or chemical attack on a fiberglass-reinforced plastic will begin at the glass-resin interface. Therefore, the glass fibers must not "bloom" or raise above the resin surface. To prevent blooming and exposure of the glass-resin interface, a resin-rich surface is necessary. This resin-rich surface is produced by an unreinforced gel coat or through the medium of a surfacing mat.

Surfacing mat is a very thin and highly porous mat of monofilaments preferably of type C glass. The filaments of the mat are treated with a silane size and bonded with a polyester emulsion. Such a mat is applied at the surface of the reinforced plastic. It improves surface appearance and resistance to weathering and chemicals. By drawing a slight excess of resin to the surface it covers surface irregularities. This type of mat therefore is useful for lamp shades, trays, and building panels.

Since the surfacing mat is designed to provide a resin-rich surface, it should not be considered as a reinforcing mat.

QUESTIONS

1. How would you formulate an electrically conductive plastic?

2. If air voids cannot be avoided in a molded transparent plastic, what is the best way to hide them?

3. Why must a solvent have a small molecule?

4. (a) What is a chlorinated solvent? (b) What are its advantages over standard hydrocarbon solvents?

5. What is the purpose of a plasticizer?

6. Why are plasticizers not used in PTFE (Teflon)?

7. Why are plasticizers not favored in plastics exposed to the weather?

8. Why is carbon black blended into polyethylene pipe?

9. What is the difference between a dye and a pigment?

10. What are the reasons for using fillers in vinyl floor tile?

11. Are dyes or pigments used in PMMA? Why?

12. What purpose is served by adding the following fillers to a plastic formulation? (a) zinc oxide; (b) antimony oxide; (c) clay; (d) mica.

13. Why is a protective sizing required on glass reinforcing filaments?

14. What is the function of a coupling agent on glass reinforcement?

15. State the effect of increasing reinforcement on the following properties of a reinforced plastic: (a) stiffness; (b) tensile strength; (c) ductility; (d) creep.

16. What weave, yarn, or fiber would you select for the following reinforced plastic products? (a) A thermoplastic office machine housing; (b) a boat; (c) a fishing rod; (d) a well casing (pipe).

17. Explain how surfacing mat protects a glass-reinforced polyester.

18. Figure 6.12 shows the condition of maximum packing of fiber reinforce-

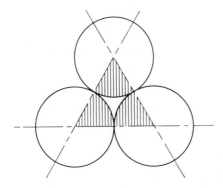

FIGURE 6.12 — Condition of maximum packing of cylindrical reinforcing fibers in thermosetting resin.

ment in a plastic. What is the percentage of reinforcing and of resin for this ultimate condition?

19. What type of reinforcing gives isotropic and anisotropic strength in a reinforced plastic?

INVESTIGATIONS

1. Find a method of painting UPVC. The paint must adhere.

2. Expose a sheet of rigid polyurethane foam insulation to sunlight for not less than 10 days. Note the surface effect. How deep does the effect penetrate during this period? Compare with the same polyurethane foam protected with a thin coat of aluminum paint. Is this foam opaque in thin sections?

3. Cut up some polyurethane insulating foam into small pieces with a knife. Observe the static electricity generated.

7

plastic foams

[7.1.]
characteristics of plastic foams

Plastic and elastomer foams may be considered as plastic-filler composites, the filler being "particles" or cells of gas. Such particles do not reinforce; therefore, the mechanical properties of the plastic are reduced. There is less plastic, because of the gas voids, so strength, stiffness, and impact strength are lowered.

Though wood is not foamed, it is a cellular polymeric material, the cells containing both air and water. Its mechanical properties, especially strength and stiffness, exceed those of any synthetic polymer foam, or indeed any unreinforced polymer. Various species of wood give a range of specific weights from 10 to 60 pounds per cubic foot; this range almost spans the range of weights available in the foamed plastics. Wood is anisotropic, with best mechanical properties parallel to the grain, the direction of the long axes of the wood cells. The synthetic foams likewise are anisotropic if the foam rises in one direction; the best properties are parallel to the direction of foam rise.

There are few plastics that have not been foamed commercially. Plastic foam technology has seen rapid advances in only a few years, so that even special areas such as rigid polyurethane foam have become extensive technologies. It must be emphasized that a foamed plastic is usually totally unlike the same plastic in the solid: unlike in properties, in processing, and usually in applications. The most remarkable example of such dissimilarity is urea-formaldehyde. The solid UF is one of the hardest plastics; UF foam is the softest of all the forms of plastics.

Two types of foams are produced: open cells and closed cells. In an open-celled foam the gas cells are interconnected with each other. In a closed-cell foam each gas cell is totally enclosed by thin plastic walls. Either type of foam may be possible in some plastics, polyurethane, for example. An open-celled structure will permit the passage and absorption of water through the minute interconnecting channels of the cells, while only water vapor can be transmitted through the integral walls of a closed foam. In practical terms, an open-celled foam would not be selected for a floating dock, a surfboard, a ski, or a life preserver ring.

The applications of foam fall into three broad types:

1. Insulating foam.

2. Cushioning foam.

3. Structural foam.

Insulating foam is a very lightweight foam used for heat insulation of buildings and cold storage areas; it weighs about 2 lb per cubic foot. The insulating properties are primarily obtained from the gas cells of the foam, hence the necessity for light weight. Cushioning foam is soft and resilient and finds uses in carpet underlay, weatherstripping, seat cushions and upholstery, and packaging. Structural foams are stronger and stiffer than the other types and therefore are capable of supporting stresses. Structural foams have a dense, hard skin of minimal foaming encasing a lighter foamed core (Fig. 7.3). The hard skin is produced by foaming against a cold mold surface that suppresses expansion of the foaming gas; the gas in the interior of the part is not cooled and expands more freely.

[7.2.]
the foaming process

One of the attractions of foam technology is its great versatility. The integral skin just mentioned is one technique. If a hot mold surface is used, then an integral skin is not produced, giving a more uniform density throughout the product. The density of the foam may be altered by changing the blowing agent, or by injecting more or less plastic into the mold in order to create a higher or lower mold pressure, and a resultant higher or lower density of foam.

Foams are produced by the familiar processes of injection molding and extrusion, or by pouring in place or spraying. The foaming action must, however, occur at exactly the right moment in the process. The viscosity of the plastic, which is controlled by the temperature, must be neither too high nor too low at the moment when the foaming gas begins its expansion, so that the gas will be contained in minute cells of the required size. Clearly, under the wrong conditions it would be possible to blow a single large bubble within the plastic or at its surface. Surface-active agents (surfactants) may be added to the foam formulation to control cell size. Polyethylene is often cross-linked in order to strengthen the cell wall against rupture from the pressure of the gas.

The foaming agent is a gas or a volatile liquid with a boiling point below 110° C at atmospheric pressure. The liquid foaming agents are usually paraffin hydrocarbons such as pentane or hexane, or halogenated hydrocarbons such as methyl chloride or Freons.

The foaming agent may also be produced by chemical decomposition of a material by the influence of heat. The decomposition must occur in a narrow temperature range at which the thermoplastic resin has the proper viscosity for foaming. Compounds such as azobisformamide which release nitrogen are particularly useful in thermoplastic foam technology. The decomposition temperature of azobisformamide can be regulated from 160 to 200° C by the use of activators and other additives.

[7.3.]
polystyrene
foam

This is the most widely used rigid plastic foam. It is made by either of two methods:

FIGURE 7.1 — Several plastic foams. The three upper foams are polyurethane on the left, polyethylene in the middle (feathery appearance), and urea-formaldehyde on the right (snowy appearance). The small dark pieces are cross-sections of a foamed PVC molding. Polystyrene beadboard is on the bottom.

FIGURE 7.2 — Packaging for a slide projector made of expanded polystyrene beads.

1.   By expanding polystyrene beads to fill a mold.

2.   By extrusion.

The bead molding method uses steam to heat the mold and thus expand the beads. Beads of various sizes are used; foamed drinking cups use very small beads, whereas insulation boards (beadboard) use large beads. Densities from 0.8 to 10 pounds per cubic foot are produced. The bead structure is apparent in the finished product. In the extrusion method the mix of polystyrene, blowing agent, and additives is extruded through a die and expanded.

Expanded polystyrene is rigid, with good resistance to water vapor transmission. The lightweight insulation boards have low thermal conductivity, K = 0.24 Btuh, though polyurethane insulation is much superior. Polystyrene has poor resistance to solvents, and many solvent types of adhesives cannot be used to bond this foam. It is not resistant to ultraviolet radiation, but can be pigmented for color and ultraviolet protection.

[7.4.]
PVC foam

Figure 7.3 is a cross section of an extruded high-density PVC ceiling cove molding. Rigid high-density foamed PVC is chiefly extruded into moldings, baseboard, window and door trim and frames, fence posts, fence palings, shelves, house siding, door sills, and pipe. Most of these applications compete with wood. In finished form such extruded shapes may be difficult to distinguish from wood when stained a dark brown. Even the density of these foams corresponds to that of wood—about 30 pounds per cubic foot. High-density foamed PVC, however, is inferior to wood in two properties. The modulus of elasticity of such extrusions is about a fifth that of wood—about 225,000 against 10^6 or more for wood. Therefore, for window and door frames sufficient cross section is required for stiffness. Also, while wood has a negligible thermal expansion parallel to the grain, these wood substitutes have the usual high thermal expansion of thermoplastic, about 0.00005 in./in. $-°$ F.

ABS foam has a specific gravity of 0.5 to almost 1.0, compared with 1.04, the specific gravity of the bulk plastic. This foam is either injection-molded or expansion-cast. The injection-molded product is used for furniture, drawers, picture frames, wall plaques, and related applications. A hard skin encloses the foam, making the product tough, hard, and abrasion resistant. The screw-holding strength of this foam is comparable to that of hardwoods.

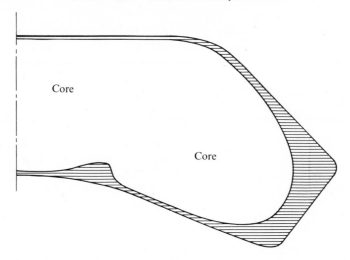

FIGURE 7.3 — Cross-section of foamed PVC cove molding showing the shape of the higher density integral skin of this extrusion. Only half of the symmetrical molding is shown.

[7.5.]
polyethylene foam

This foam has a soft, feathery appearance in low-density versions. It is available in densities from 2 to 9 lb/cu ft. Like bulk polyethylene, the foam is very tough and rubbery, with excellent chemical resistance. Water-vapor transmission is low. Resistance to ultraviolet radiation is superior to that of the bulk plastic, though the foam is unsuited to indefinite exposure to ultraviolet. Unlike bulk polyethylene, the foam can be bonded with epoxy adhesives.

The characteristics of polyethylene foam make it a versatile material. Uses include flotation applications in water sports equipment, padding for baseball, wrestling, and other sports, cushion packaging, weatherstripping and gaskets, and construction joints in buildings. Its K-factor, 0.35 Btuh, is too high for it to be used as building insulation, and so is its cost.

[7.6.]
urea-formaldehyde foam

Urea-formaldehyde is the lightest of the insulating foams, weighing only 0.7 pounds per cubic foot. Its K-factor is 0.2 Btuh/in. It has virtually no mechanical strength even in compression, is white in color, and does not support combustion. About 60 percent of the cells of foamed urea-formaldehyde are open, and the foam can absorb about 30 percent water by volume.

Like polyurethane foam, urea-formaldehyde is foamed on the jobsite. The foaming system has two components. One is a water solution of urea-formaldehyde and the other a water solution of the foaming agent and a catalyst for curing the resin. The material does not bond to surfaces as polyurethane foam does but will flow into crevices and cavity walls before curing. Unlike polyurethane foam, it cures with little exotherm. The foam dries out over a period of time to an extremely soft condition somewhat resembling wind-driven snow.

[7.7.]
rigid polyurethane foam

Polyurethane foam planks are available in densities of 2, 3, 4, 6, 8, and 10 pounds per cubic foot. These are produced by foaming a very large volume of material and sawing the large "bun" into planks on a bandsaw. Such planks have the same applications as polystyrene foam planks. They are, however, not quite so brittle, are stronger, are not attacked by most solvents, and, being thermosetting, will withstand higher temperature than foamed PS. The two-pound density is usually installed for insulation, but four-pound foam is more resistant to collapse of the foam by heat, cold, or ice formation in roofs.

Low-density polyurethane foams are chiefly used as heat-insulating materials. Artificial limbs use foams in the range of 12 to 18 lb/cu ft. These densities and higher, up to 40 lb/cu ft, are cast into furniture components and picture frames. An integral skin can be formed on such molded products.

Polyurethane foams, rigid or flexible, are produced by a reaction between a diisocyanate such as tolylene diisocyanate (TDI) and an alcohol or amine. A small amount of water is used in the reaction in order to form carbon dioxide. This CO_2 is the foaming agent, though Freons may be used in foams of low density. In cold weather, Freons may have to be added to obtain sufficient foam expansion. A silicone oil is included in the formulation to keep the cell size small. The polymerization reaction produces linkages of urea with ethane, hence the name polyurethane. The density of the foam is determined by the type of polyester used and the amount of water, about three percent, which controls the amount of CO_2 generated.

The gas in the closed polyurethane cells is, therefore, either CO_2 or a Freon. The thermal conductivities of these gases and of air are these:

FIGURE 7.4 — A cupboard door of poured polyurethane foam. The foam density is 16 lb/cu ft, but mold pressure produces a final density of 22 lb. The increased density is required for screw and nail holding strength. Both the color and grain of red oak are simulated.

Freon-11	0.058 Btuh/in.
CO_2	0.102 Btuh/in.
Air	0.168 Btuh/in.

The Freon-blown foams are the best insulation. Both CO_2 and Freon produce a superior insulating foam to polystyrene foam. However, any comparisons must take account of an aging process that may occur in polyurethane foam. Freon-foaming polyurethane in a density of 2 lb/cu ft has an initial K-factor of 0.11, far superior to any other insulation. If, however, this insulation has access to air at its surfaces, air will permeate through the cell walls and replace the Freon. This exchange of gases raises the K-factor to 0.16 Btuh, which is still superior to polystyrene foam. If the foam is enclosed so that air does not have access to it, the K-factor should remain at its initial low level.

Polyurethane foams are stronger in the direction of foam rise than in the cross-direction. Depending on which direction requires the better strength, panels may be foamed in the vertical or horizontal position. Significant properties of polyurethane foams as a function of foam density are given in Figs. 7.5 through 7.7. These properties are measured in the direction of foam rise. They are average values, since different urethane formulations give some difference in values. Mechanical properties such as modulus of elasticity, ten-

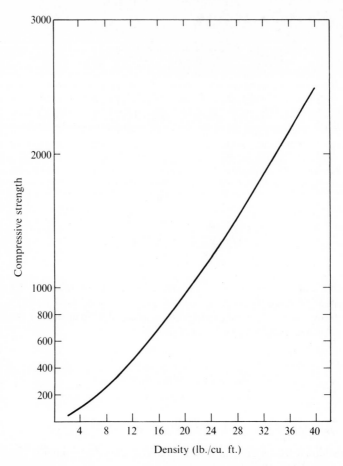

FIGURE 7.5 — Average values of compressive strength of rigid polyurethane foams.

sion, compression, and shear strength are not proportional to density but increase rapidly with increase in density.

[7.8.]
installation of polyurethane
foam

For standard spray contracting, such as insulating the walls and roofs of
buildings, a complete spray installation, truck-mounted, is necessary. The two

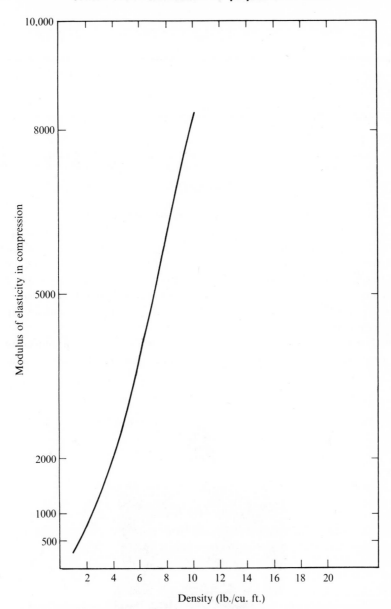

FIGURE 7.6 — Average values for modulus of elasticity of polyurethane foams. Specific types may vary somewhat from these values.

resin components are pumped from 55-gallon drums or larger containers and then mixed, heated, and discharged from a special spray gun. The two components react very rapidly and therefore must not mix until discharged from

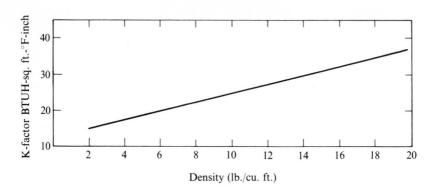

FIGURE 7.7 — K-factors for polyurethane foams.

the gun. If they become mixed in the hose, heater, or gun, there is no solvent that can remove the cured plastic. For on-site spraying, the two components are mixed in equal volumes, though factory application of polyurethanes may be made using other proportions. The sprayed liquid foams immediately as it is deposited on a surface and is hard within a minute. A complete spray installation for contracting is shown in Fig. 7.8.

For the manufacture of furniture components or other products under factory conditions using higher density foams, equipment is more complex.

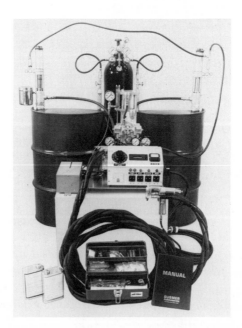

FIGURE 7.8 — Polyurethane foam spray installation.

Such products are made either by spraying, pouring, or frothing the resin. In frothing practice, the resin is partially foamed before discharge, the foaming being completed after discharge from the nozzle. The small aerosol kit of Fig. 7.9 uses froth.

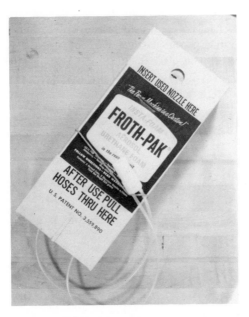

FIGURE 7.9 — Small aerosol kits of polyurethane foam are available, such as the one shown here. It supplies 9 board feet of foam as a pre-expanded froth.

[7.9.]
flexible polyurethane foams

The flexible urethane foams are open-celled. They are produced by mixing an emulsion of a polyether with tolylene diisocyanate and various catalysts. As with rigid foams, the amount of carbon dioxide generated for foaming is determined by the proportion of water, more water giving a lighter foam.

Typically these resilient foams will have an elongation of 200 to 225 percent, which is lower than the elongations for solid rubbers generally. Such flexible foams if compressed 50 percent for 24 hours will fully recover this strain when the load is removed.

Major applications are furniture, upholstery, and carpet underlays.

[7.10.]
other foams

Many other plastic foams have more limited use. Phenol-formaldehyde foams, like UF foams, are of low strength and brittle. Polycarbonate, polyphenylene oxide, nylon, and other thermoplastic foams are injection-molded. Some of these can be reinforced with glass fiber, for improved heat resistance, impact resistance, strength, and stiffness (approaching that of wood). Such reinforced foam is well adapted to large lightweight structures such as business machine housings, shrouds, doors, panels, bodies of recreational vehicles, ducting, and tote boxes.

QUESTIONS

1.   Why do some plastic foams have anisotropic properties?

2.   Differentiate between open cells and closed cells.

3.   Epoxy cement will bond to a cut edge of PE foam but not to solid PE. Explain why.

4.   Which of the foams will flow before setting up?

5.   What advantages does polyurethane foam offer over styrofoam?

6.   Explain how the K-factor of urethane foam can increase from an initial 0.1 Btuh to about 0.15.

7.   What effect does increasing foam density have on the following properties of the foam?
   (a) compressive strength   (b) stiffness
   (c) thermal conductivity   (d) nail-holding strength

8.   Urethane foam has two components, each component being pumped from a 55-gallon drum. When a pair of drums are emptied, two full drums are substituted, but unfortunately the operator interchanges the pumps. Explain why two new pumps must be bought.

INVESTIGATIONS

1.   Compare bead size in a styrofoam coffee cup with that in a construction beadboard.

2.   Test various adhesives, including model aircraft cement, for their suitability in bonding styrofoam.

PRACTICE

8

design
in plastics

In general, each of the plastics is widely different from other plastic materials both in properties and applications. Phenol-formaldehyde and UF may be somewhat similar in properties, but UF can be colored whereas PF cannot, and this difference has a very marked influence on applications. Cellophane, polyethylene, cellulose acetate, and polyvinyl fluoride are all excellent film materials. But cellophane and polyethylene are unsuitable for photographic film, acetate is not suitable for outdoor exposure to the weather, PVF and PE do not have the clarity of the other films, and so on. Before anyone can understand, use, or design in plastics, he must have samples of the material to examine and to test, and he must have information on properties.

Perhaps there is no better way to become assured in the management and use of the plastics and elastomers than to practice designing with these materials. By attempting designs, and if possible also making up the product thus designed, you quickly obtain a thorough understanding of the properties, the manufacturing problems, and the advantages and disadvantages of the polymers. A number of products are suggested in this chapter for which the reader may attempt an original design. It is unimportant if the reader's designs are imperfect; the purpose here is to obtain a competent understanding of the plastics. If, in addition, the reader obtains some ability in industrial design, then that is an additional benefit. But designing products is always a matter of many complications, so for those unfamiliar with design work, suggestions and hints, and where necessary warnings of failure, are supplied. Mock-ups of the designs should be made whenever it is possible in order to obtain as much practical experience as the circumstances will allow.

Successful design using plastic materials requires a knowledge of both the composition and character of the materials and the methods of shaping plastics.

Compared with metals, the plastics are complex materials. Most plastic formulations are mixtures of materials, with both organic and inorganic materials being used. Even a pure resin is a mixture of molecular weights, and it may have both amorphous and crystalline regions.

Metal products can often be shaped by machining methods using a plate or a rod as stock. While this is sometimes done with plastics, nevertheless machining is usually too expensive a shaping method for these materials. The designer must have a reasonably clear notion of the feasible methods of producing the shape he requires as well as a feasible choice of plastic material. A duck decoy, for example, may be made of solid plastic in a hollow shape or of solid foam of the same plastic. In deciding between the two methods, the designer must know that the production methods are entirely different, even though the same plastic is used. There are still inventors who spend over a thousand dollars to patent an idea, only to tell the prospective manufacturer that "it should be made of plastic". This is perhaps the most uninformative statement that anyone can make, and it indicates that the inventor has a wholly incomplete invention. Until the plastic and the manufacturing method are decided, no one knows whether the product cost will be 10¢ or $10, or whether the mold cost will be $100 or $100,000—though the inventor at this stage of his invention wants a dollar royalty on each item produced.

The Patent Office records show many inventions in plastic that present formidable manufacturing problems. One is an extrusion that can't be extruded successfully. Another is a hollow, tapered thermoplastic part with a length 50 times its width, a shape that approaches the impossible for manufacturing, yet is required in a consumer product that must be manufactured for not more than $3–$4. Designs should be made whenever it is possible in order to obtain as much practical experience as the circumstances will allow.

[8.1.]
selecting plastics for a product

The remarks of Sec. 1.2 apply to the selection of plastics and plastic formulations for product applications. The several requirements imposed on the material were stated to be

1. Service requirements.
2. Manufacturing requirements.
3. Economic requirements.

In the initial stages of designing a product, service requirements are the chief preoccupation. After the design is roughed out, modifications can then be incorporated to make it more convenient for manufacture. The several plastics converting methods are discussed in following chapters, and these must be consulted for details that govern the production method selected.

The plastic article must be produced within a suitable cost range. This product cost includes material cost, manufacturing cost, and various overhead items such as administration and marketing. The most powerful influence on costs is always volume of sales. If, for example, a $20,000 molding die produces only 20,000 pieces, then the die charge is $1.00 per piece; if two million pieces are produced, the die charge falls to 1¢ and product cost drops by 99¢. Hence, the importance of a sufficient market before a product is designed.

Costs are an enormously complex subject to discuss, since both material and manufacturing costs are involved. The objective in costing a product is the minimum overall cost. Sometimes a less expensive material may have adverse manufacturing costs that cancel out the savings of a cheaper material, or may have an excessively high scrap rate. But since costs of materials and of manufacturing them are continually changing, vary from company to company and location to location, and are subject to such variables as transportation, storage, and overhead items, it is clearly impractical to make useful cost analyses in this book, and we shall not attempt to do so.

[8.2.]
solving design problems

There are two methods of working toward a solution to a problem.

One method, the linear method, was taught to you in many school courses, such as mathematics. You are given certain necessary information, and with a series of steps or operations you proceed logically from this given information to the one and only answer. Unfortunately, this method can be used only in very artificial circumstances. With most problems in life, including technical problems, the problem is either wrongly defined or not clearly defined, the necessary preliminary information is incomplete or uncertain, and the solution cannot be obtained by logical operations.

Particularly in design work it is often much easier to work the problem backwards. This is the cyclical method. To solve the problem, first get a solution to it. Any possible solution will do. Then analyze this solution to see if it is acceptable. If not, adjust it to obtain a solution that conforms better to the requirements. Keep improving the solution until you decide that you have a solution that is finally acceptable. You can obtain these solutions by many methods: by logic, by brainstorming, by criticism, by creative inspiration, by asking other people, by doodling.

To illustrate the cyclical method, consider once again the plastic hockey stick. What are the requirements? After some thought, you decide that toughness and resistance to breakage are the primary requirements for the blade. You leaf through plastics catalogues to find a tough plastic, and select polycarbonate or cellulose propionate for the blade. For the handle you select a rigid foamed high-density high-impact grade of extruded PVC, because this material is cheap and also approximates the density of wood. You now have a tentative solution. You can now prototype such a hockey stick to test your first solution. This solution is not satisfactory, one of the reasons being excessive flex in the handle. You have now to solve the flex problem, which will lead you into a more successful design. By continued modification in the light of tests, you will in time arrive at a good design. Indeed, the design process never ends, because there is always an improved answer to be found. But since you cannot redesign forever, you must at some stage fix the solution and abide by it.

Charles Kettering of General Motors once said, "Every great improvement has come after repeated failures; virtually nothing comes out right the first time."

[8.3.]

a procedure for design

The foregoing remarks suggest that a design project should have five stages or steps:

1. The statement of the problem.

2. Analysis and research.

3. Tentative solutions.

4. Experiment and testing.

5. A final and accepted solution.

1. the statement of the problem — Before you can find an answer, you must know what the problem is. There must be a clear and direct statement of the design requirements. In any design work, it is well to stop occasionally to ask, "What, exactly, are we trying to do? Are we solving the wrong problem?" Very often you do not know what the problem is until you are deep into it.

The following is a poor example of problem definition: Design a plastic hockey stick. Obviously any polymer material, thermoplastic or thermosetting, can be given the shape of a hockey stick. This definition omits much information needed by the designer. What price range? What quality? What features? A stick for children or for professional hockey players? Should the stick be unbreakable (or nearly so)? Are patentable features desired in the design? The designer can be successful only when he knows within limits what is expected in the design.

2. analysis and research — When the design problem is set, there is a groping for information from which to obtain some design concept. When he begins, the designer may not be able to find a starting point or starting idea. Time, therefore, must be spent in reflection. Catalogues, technical literature, or technical salesmen may be consulted for ideas. The author allows his mind to search the problem area while driving to and from work, or perhaps he stops at a coffee counter and sketches on a paper napkin at the counter. A restless and unrelenting mind can find a good basic answer to a design problem after a period of time (in the author's experience, from a week to three months). After a tentative solution is found, a further period of reflection is needed to think of any flaws or false reasoning.

3. tentative solutions and testing — As soon as a tentative solution is found, problems become defined. As Kettering infers, the tentative solution can be expected to be a failure. Consider the creative idea (tentative solution) of a hockey stick with a polycarbonate blade and PVC handle. The experienced designer would at this stage not congratulate himself on his creativity. Instead, he would reflect that the polycarbonate must be bonded to the PVC handle and that joints are usually trouble spots. He would experiment with various cements to find a reliable cement and a reliable bonding method. When and if the problem of the joint is solved, he would roughly mock-up the blade and handle, perhaps machining them by hand. This prototype would be checked out and tested by himself, and the opinions of others would be solicited. Defects would certainly be found, all of which must then be designed out of the product. Inevitably these solutions to problems lead to modifications to the design or even to the materials.

At an early stage the manufacturing problems must be solved also. The first attempts at manufacturing the product rarely go well. Voids appear in the plastic, or the product is warped, or the surface finish is unacceptable. It is not unusual to spend a few months manufacturing scrap in the effort to overcome the manufacturing problems.

Thus the road to successful product design is rather long, full of interruptions and disappointments, and fumbling for solutions to a hundred major and minor problems, most of them unexpected. The bright idea that you rush to patent is only the first step in a hundred steps, and often the easiest one. Yet it is interesting that there is always a solution to any technical problem, if only the designer has faith that there is a solution. It is usually so simple that the designer is slow in finding it. Only persistence is required for success. The greatest inventor of them all, Thomas Edison, said that genius is one percent inspiration and ninety-nine percent perspiration.

[8.4.]
product design

We define product design as follows: Product design is the use of materials in a new and creative form or organization for the purpose of serving a defined function. This definition rules out mere ornamentation. The arrangement of a duck-hunting scene on a plaque or the printing of a design on a dinner plate are not product design by this definition. Design must be built into the form of the product, not added as superficial cosmetics.

Two design principles can safely be recommended, even though there are many departures from these principles. One principle is *honesty*, the other *simplicity*. Both are easily understood.

By *honesty*, we mean that the product should reflect the advantages and essential character of the materials used. That is, the inherent qualities of the material should be exploited, so that it does not imitate another material. Though this excellent principle is often breached, it is the foundation principle for the best designs. If softwood is stained to imitate hardwood, this is not honesty, though the reason for doing so, economy, is justification. Imitation ceiling beams are made of polyurethane foam stained to look like wood. Again, this is not honesty, but the advantages of handling a light-weight "wooden" beam are powerful reasons for breaching the rule of honesty.

As for *simplicity*, there is a tendency among technical personnel to find complex solutions to problems. We tend to worship complexity, mathematical elaboration, and analysis. Although these intellectual activities certainly have their place, the ideal product design should be one so simple and straight-forward that any casual passerby would have the impression that he could have done the job. Simplicity applies to such aspects of design as clean and simple lines and layout, without the elaborate ornamentation that character-ized the domestic products of the period of a hundred years ago. The chairs in Robert E. Lee's living room would be valuable as antiques but are not saleable in today's furniture stores. Many such elegant product designs were not even comfortable or efficient; nowadays "form follows function"—the shape must be governed by the use to which the product is put. The earliest automobiles were designed to resemble stagecoaches or horse-drawn car-riages; the modern automobile has a form determined by comfort, low center of gravity, and visibility.

color —— Except for a few applications of transparent plastics, those plastics that are molded into products are pigmented. Color has a powerful psychological impact, and because of its appeal it is an important element in design.

Insights into the use of color, both bad and good, can be obtained from examination of packages and cartons on supermarket shelves. Certain colors are not acceptable on certain products. Pink is not used on pipe wrenches because it suggests femininity. One of the associations of green is poison, and this color is not found on packages of sugar. Blue and white are suitable colors for sugar: white for purity and blue for distinction. The same colors would be suitable for a set of plastic salt and pepper shakers.

Most colors have many associations. Which association is called to mind may depend on the product that is pigmented.

Green	Poison, ill health, nature, quietness.
Blue	Distinction, coolness, passivity.
Purple	Royalty, dignity, distinction, richness.
Red	Heat, warmth, excitement, action, blood.
Yellow	Warmth, sunshine, ill health.
Orange	Warmth, warning.

White   Purity, formality, cleanliness.
Black   Death, gloom, evil, night, severity, formality, religion, distinction.

Black properly used can be the most outstanding of all colors (it is used in women's evening gowns) and is excellent as a background or trim for bright colors. Otherwise, black, gray, and brown are negative or neutral colors.

The effect of a color depends on its hue or tint. One yellow may be sickly, another brilliant. A certain green in a certain application may suggest poison, while a different green may suggest nature and trees.

These remarks give only basic approaches to product design in plastics. Enough design projects are now suggested so that the reader can select those that best motivate him or her to creative effort in plastics materials. In every case the color or color design should be selected for the product to be designed.

[8.5.]
a rack for phonograph records

the design problem — A holder for 12-in. phonograph records, approximately 12 inches long, preferably of modular design. The design should be a luxury model for music lovers who value their recordings and will pay extra for a well-styled, striking holder.

a possible solution — Suppose the designer produces the sketch of Fig. 8.1. He is attracted to $\frac{1}{8}$-in. PMMA for the partitions, in a dark smoky gray or blue-gray. For the bottom and back, which comprise the frame, the designer wants a material that contrasts in color and texture with the PMMA.

Any type of plastic sheet material will have a similar smooth surface to the PMMA and will be inconvenient to join with the PMMA dividers. How about foam 1 inch thick? This can be grooved to receive the dividers. Polyurethane foam is dusty in low densities, and is dark in color. Polystyrene foam is fragile and friable.

Polyethylene foam is white, which contrasts well with the dark dividers. Its soft surface gives a contrasting texture to the PMMA sheet.

But adhesives do not bond to PE. Perhaps they will to PE foam because of its porosity. A simple test shows that epoxy adhesive bonds well to

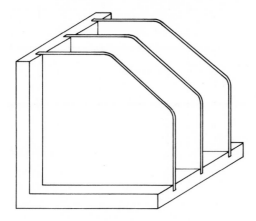

FIGURE 8.1 — A preliminary sketch for a phonograph record rack.

the foam. The base and back can be made of one piece of foam. There is no difficulty in bending the foam where back and base meet, since the foam is compressible.

The shape of the dividers must be an attractive curve properly proportioned.

There are some possible objections to this design. If the foam becomes dirty, is it cleanable? If exposed to ultraviolet radiation, will it turn yellow?

[8.6.]
a wall panel

the design problem — A wall panel 4 feet wide by 8 feet high using contrasts in color and texture. Either use plastics materials exclusively or mix plastics with wood, metal, or textiles for contrast. Suitable metal sheet materials are stainless steel, copper, and aluminum. Do not be concerned with cost or manufacturing method.

suggestions — The problem here is that of contrasting areas, colors, and textures. The areas may be separated by lines, horizontal or vertical. (Diagonal?) The various areas must combine in a design that has some unity (that is, not a jumble of mixed and unrelated areas), some variety, with proportion and balance. Successful variety could be obtained, for example, by contrasting black plastic with stainless steel.

Some suggestions for good design are given in Fig. 8.2.

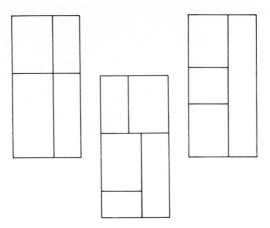

FIGURE 8.2 — Some preliminary sketches for decorative wall panels.

[8.7.]
a plastic beehive

Figure 8.3 shows the dimensions of a standard beehive, called a "honey super." The honey supers are stacked one above the other, the bottom one having a bottom piece under it and the top piece being fitted with a cover to protect the stack against the weather. The honeycombs are hung in the supers from the shelf rabbeted in the sides of the honey super.

Honey supers are made of wood. The bee colony can maintain climate control in the beehive provided the beehive is not wet. Wood, however, readily absorbs moisture.

Consider the possibility of a honey super made of sprayed insulating rigid polyurethane foam, perhaps 4 lb/cu ft. The honey super could be made of four slabs joined to make the box, or the four sides could be sprayed in one piece integrally. If sprayed in one piece over an interior form, there will be some overspray, and it would be best to make a honey super perhaps six feet long and then to cut this long length into standard lengths on a bandsaw. The form will require some kind of mold release because polyurethane is quite adhesive. Would a stretchable cloth painted with silicone RTV rubber and pulled over a plywood form serve the purpose?

Invent a honey super made of urethane. Provide all dimensions, details, and manufacturing information. If possible, make a prototype. Expose a piece of urethane to outdoor sunlight for a considerable period of time to determine whether it requires ultraviolet protection (the change of color due to ultraviolet exposure probably is not important for this product).

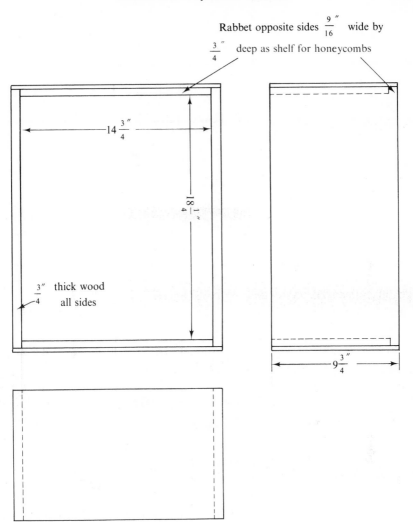

Rabbet opposite sides $\frac{9}{16}''$ wide by $\frac{3}{4}''$ deep as shelf for honeycombs

$14\frac{3}{4}''$

$18\frac{1}{4}''$

$\frac{3}{4}''$ thick wood all sides

$9\frac{3}{4}''$

FIGURE 8.3 — A "honey super." These are stacked to make beehives. The honey super unit has no top or bottom—only four walls.

[8.8.]
synthetic cork tile

Cork floor tile is very attractive in appearance and makes an excellent resilient floor. It is quite expensive, however. Can you invent an imitation cork tile using solid polyurethane foam or pieces of such foam bonded together? You must obtain the same dark brown color of the real cork.

[8.9.]
duck and goose decoys

Existing duck decoys do not provide a feathery texture on the surface of the decoy. One of the common plastic foams has a feathery appearance, especially at a cut surface. Could duck decoys be foamed from this plastic in a closed mold? Forming in a closed mold will tend to produce a smooth surface, but perhaps the surface could be given a feathery texture by passing a high-speed grinder or cutter over the surface.

[8.10.]
salt and pepper shakers

Design a set of salt and pepper shakers in plastic. Make sure the design is simple and clean. Avoid cute designs—form must follow function. Make sure that the shaker can be manufactured easily. Select a color scheme. Figure 8.4 offers some suggestions for styling.

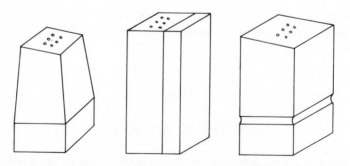

FIGURE 8.4 — Sketches for the design of salt and pepper shakers.

[8.11.]
a drafting board of polymer materials

Invent a 24 × 36-in. drafting board made solely or chiefly of plastics and elastomers.

List first the requirements of an ideal board and try to design from these requirements.

A drafting board is simply a surface, or two opposite surfaces, of appropriate properties, supported by a core material that will provide stiffness and resistance to warping. You will require, among other things, a nonstaining and easily cleaned surface. Here are two suggestions:

1. Ionomer film (what thickness?) on wood veneer on polyurethane foam (what density?). To prevent warping, the board should have a symmetrical construction; that is, both sides of the board should be of ionomer-wood veneer.

Assume that the foam is self-bonding if poured against the wood veneer. The adequacy and permanence of the bond would be proved by testing.

2. Polyvinyl fluoride film painted on the underside with chlorosulfonated polyethylene (Hypalon). The purpose of the Hypalon is simply to provide a beautiful appearance, and may be colored. The core of the board will be polyurethane. Between the urethane and the PVF-Hypalon surface there must be a harder material (a rubber?) to resist the penetration of compass points. Since Hypalon is dissolved in a solvent, evaporation of the solvent after painting may lead to shrinkage effects, which may or may not be a source of problems.

Any design also has the problems of the four edges and the manufacturing method. Will the boards be made in final 24 × 36-in. size or in a large area to be cut into 24 × 36-in. pieces?

[8.12.]
a meeting minder

The meeting minder was designed by a man who decided to do something about the appalling waste of time represented by meetings. It is sketched in

Fig. 8.5, and its purpose and method should be self-evident. Design a housing for the meeting minder, selecting a plastic and color scheme.

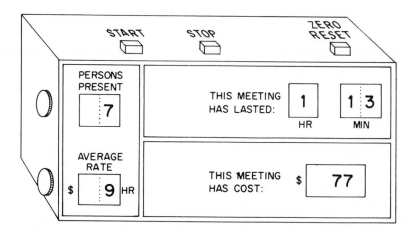

FIGURE 8.5 — A "meeting minder" to remind personnel of the cost of talk and discussion (invention of R. Blicq).

[8.13.]
a fish cooler

Design a portable cooler suitable for fishermen to carry home their catch, not less than 36 in. long. Include an ice compartment, and provide a lid, upholstered or otherwise, suitable for a seat in a boat or at a picnic. Typical coolers may be seen in sports departments of large stores.

[8.14.]
coasters

How might you produce coasters (the pads placed under drinking glasses) cheaply from plastics? Presumably you will imprint or impose a decorative pattern.

[8.15.]
plastic eavestroughing
for houses

Plastic rain gutters are more commonly used in Europe than in America. An acceptable design of eavestroughing must meet the following requirements:

1. They must be easy to manufacture.

2. They must be convenient in shape for packaging and transport.

3. They must be easy to erect.

4. There must be no problems with the high thermal expansion of plastics.

5. They must be paintable to match any color scheme of a house.

6. They must weather well and resist ultraviolet radiation.

7. They must meet the temperature and sunlight extremes of Louisiana and Montana.

[8.16.]
a plastic hockey stick

Figure 8.6 shows the dimensions of a hockey stick. Invent a plastic hockey stick of professional quality for mass production. The following are some suggestions to consider or reject.

The blade must be able to take punishment—should it be acetal, polycarbonate, nylon, cellulose propionate? How about a suitable density of polyurethane foam reinforced with surfacing mat? Note that the thickness of the blade shown in the figure cannot be successfully injection-molded except in a foamed plastic, such as foamed polycarbonate.

If you join the blade and the handle by adhesive bonding, be warned that no adhesive joint design is reliable until proved so. An adhesive joint is always weak in peel strength—see Fig. 16.1. Protect it against peel failure by rivets or some other method.

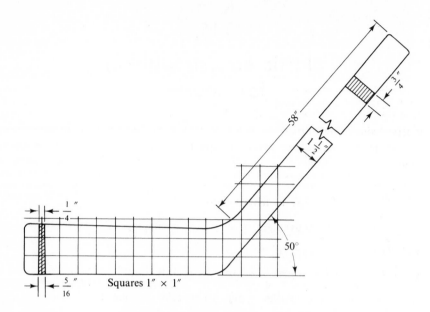

FIGURE 8.6 — Wooden hockey stick.

[8.17.]
more products for design
in plastics

1. A small convenience table, say 12 × 18 in., with one leg, suitable for holding a cup of coffee beside a chair or for taking notes at a conference. The leg and feet must be of steel tubing. Some edge designs are given in Fig. 8.7.

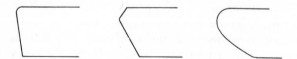

FIGURE 8.7 — Edge treatments suitable for trays and tables.

2. A dog kennel.

3. A bird house.

4. A street litter bin.

5. A tool caddy with handle and drawers for small parts.

6. A Christmas tree stand.

7. A wall clock face.

8. A waste paper basket.

9. A draftsman's pencil holder for a number of pencils.

10. Figure 8.8 shows a paint brush made of flexible polyurethane foam instead of fiber for the brush. Design a handle and a method of attaching such a brush to the handle.

FIGURE 8.8 — Paint brushes of flexible polyurethane foam. The handle is a hollow section of polystyrene.

11. A billiard cue.

9

reinforced plastics

[9.1.]
the polyester formulation

Of all the thermosetting plastics, the polyesters offer the greatest diversity of formulations, properties, molding methods, and applications. Many of the molding methods to be discussed here are equally applicable to other thermosets besides polyesters, though polyesters will be emphasized.

The polyester resin must be formulated to suit the molding method to be employed as well as the service conditions that the product must meet. The same formulation cannot be used for different curing temperatures, for example. The supplier of the resin must also know whether it is to be a coating resin, a gun-sprayed resin, or a gel coat. The different methods and uses each require a suitable gel time, and this is controlled by the resin material, monomers, promoters, inhibitors, and catalysts. For the case of a coating resin, for example, a small amount of silicone will be added to control surface tension and produce a smooth coating. If the polyester must cure in the presence of atmospheric oxygen, then a special wax must be incorporated or monomers must be selected to cure to a tackfree surface, since oxygen tends to produce a tacky surface. In molding between match-die surfaces, oxygen would not present this problem. Methacrylate and styrene are unsuited to an air cure, since they will give a tacky surface; however, these monomers are suited to matched die molding. The acrylic monomers give excellent weather resistance, but increase the mold shrinkage somewhat.

Styrene is the cheapest of the monomers used in polyester formulations. It does not weather well, and has a high rate of evaporation. Evaporation loss from an unsealed container can adversely affect the required cross-linking and may cause sticking in the mold because of incomplete curing.

The filler as well as the resin has its influence on the molding process. The use of thermally conductive fillers can reduce the cycle time considerably by influencing the cooling rate. Fillers, however, tend to absorb moisture, and such moisture content must be carefully controlled. Gel time is shortened and shrinkage increased by any moisture present in the filler material. Clay fillers are especially difficult to dry thoroughly.

[9.2.]
open and closed mold processes

A closed mold process requires both a male and a female die to form the reinforced article. An open mold uses only a male or a female die but not both.

An open mold will supply a controlled surface finish only on one side of the molded article. Surface finish and detail can be provided on both sides of a closed mold. The closed mold is impractical for very large moldings such as boat and ship hulls, but is suited to quantity production of smaller articles because of fast production rates. Though mold costs are much higher for closed molds, labor costs are significantly lower.

The principal open and closed molding processes are these:

Open molding	*Closed molding*
1. Hand lay-up	*1.* Preform molding
2. Spray-up	*2.* Mat and fabric molding

3.   Vacuum or pressure bag

4.   Filament winding

5.   Various other methods, such as encapsulation, centrifugal casting

3.   Premix molding

4.   Injection molding

5.   Continuous lamination

[9.3.]
hand lay-up

Hand lay-up, or wet lay-up, as it is also called, is an open mold process. Since no pressure is used, other than rolling with a squeegee to remove entrapped air, very lightweight and simple molds can be employed for the process. Wood, plaster, epoxy, or reinforced thermosets are frequently selected for mold materials because they are easily shaped. In this process, fabric or mat is saturated with liquid resin, and the thickness of the product is built up by applying successive layers of wet fabric. Usually a special gel coat is sprayed against the mold before the layers of fabric are applied; this gel coat provides a high surface quality and a nontacky surface. Curing usually occurs at room temperature.

This is the slowest of the many processes for forming reinforced thermosets, but also the cheapest in mold costs. End uses include boats, swimming pools, trailer bodies, flat and corrugated sheets and other products of large area. There are no size restrictions on the article to be laid up, and the method allows the designer remarkable flexibility. Changes to the design are easily made by altering the inexpensive molds. But clearly the quality of the product depends on the care used by the lay-up man.

The procedure begins by placing mat or fabric over the mold. The mats are trimmed to suit. Catalyzed resin is applied to the reinforcement and rolled thoroughly to wet out the fibers. All air bubbles must be excluded. If the back surface must be dressed, a layer of cellophane can be rolled over the lay-up. The material may cure relatively slowly at room temperature, or more rapidly in an oven.

Fire hazard and the evaporation of toxic gases into the atmosphere are safety considerations in the process.

Glass fiber is not as weather-resistant as sheet glass because of the very large surface area of the filaments. In laying up building materials or other products that must be weather-resistant, the glass fibers must not become exposed. Fiber-reinforced building materials use chopped strand or swirled mat as the usual reinforcement. If insufficiently covered by polyester, the plastic surface can erode or otherwise degrade, thus exposing the resin-reinforcement bond to attack by water vapor. Such deterioration depends entirely on a thin surface of resin covering the fibers close to the surface. The usual treatment is to apply a "gel coat" to the mold surface. The gel coat is a thin unreinforced layer of resin formulated to be tough and durable, which is allowed to cure partially, that is, gel, before the reinforcement is laminated to it. The gel coat provides a resin-rich surface to protect the fibers. The surface may also incorporate a surfacing mat of tightly bound and open fibers that are richly embedded in resin.

Resins containing methyl methacrylate have excellent weathering characteristics. Methyl methacrylate is substituted for about half the styrene monomer in a weather-resistant formulation. Spraying the surface with a coat of acrylic lacquer or polyurethane also provides weather resistance. An alternative method of weather protection is a film of polyvinyl fluoride on the surface.

[9.4.]
spray-up applications

The spray-up method is a wet lay-up technique using a spray gun. Special gun-type rovings are used that can be cut and wetted easily. The polyester resin is formulated for low draining characteristic and faster gel time. Wetting of the reinforcement is better by this method, and the production rate is faster. The molds are the same as in the hand lay-up method.

The operating method of the spray gun is shown in Fig. 9.1. The continuous gun roving is chopped and discharged into two converging streams

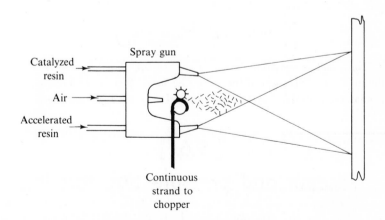

FIGURE 9.1 — Spray-up gun.

of resin. Chopped fiber can be varied in length from about ½ in. to 2 in. Separate reservoirs supply one nozzle with catalyzed resin and the other nozzle with accelerated resin. Wetting is excellent because it is performed before the fibers are deposited; therefore, fast-setting resins can be used.

[9.5.]
filament winding

Filament winding is best suited to cylindrical pressure vessels or conduit such as pressure tanks, pipe, tubing, missile fuel cells, and storage tanks, though even ladders have been made by this method. The process uses a machine that rotates a solid or collapsible core on which is wound a continuous resin-saturated roving or tape under tension (Fig. 9.2). Curing is completed in an oven or autoclave.

The filaments are wound in the direction that has the highest tension stresses.

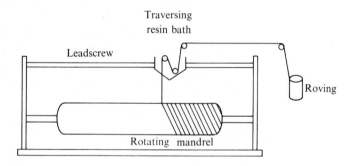

FIGURE 9.2 — Filament winding.

Hard salt is sometimes used as a core for closed vessels, where there is the problem of removal of the core. The salt is then washed out with water. An inflatable material that can be collapsed also serves the purpose.

[9.6.]
vacuum and pressure bag forming

The vacuum bag and pressure bag methods are similar; both use low pressure. In the case of vacuum, atmospheric pressure is used against the reinforced plastic. Steam or air pressure is used in the pressure bag method. See Fig. 9.3. Rubber sheeting, cellophane, cellulose acetate, or polyvinyl alcohol are used as bag materials against the lay-up. Either male or female molds are used, a male mold giving dimensional control and surface finish over inside dimensions and a female mold over outside dimensions. A hold-down ring holds the edges of the bag while a vacuum is drawn through vacuum holes in the mold.

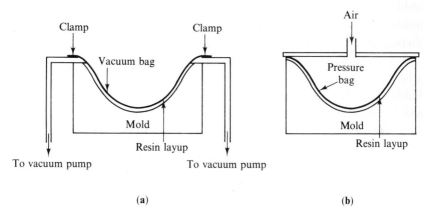

FIGURE 9.3 — Vacuum and pressure bag lay-up methods.

The application of a vacuum improves the quality of a molded part by removing air, which otherwise may produce voids, bubbles, or weak laminations.

These bag methods use standard hand or spray lay-up, but employ a small positive pressure to force the plastic against the mold.

After lay-up and bag pressurizing, the laminated or reinforced plastic may be cured at elevated temperature in an autoclave. The autoclave is a cylindrical steel pressure vessel usually employing high-pressure steam for curing reinforced parts.

[9.7.]
various open molding methods

Articles are sometimes made by casting the resin in an open mold, especially for encapsulating electrical, electronic, and similar hardware. Milled fiber or chopped strand is mixed with catalyzed resin and simply poured into the mold. Without fiber or filler, shrinkage and crazing would be intolerable in most casting of thermosets, though not for most epoxies.

In centrifugal casting, a hollow cylindrical mold is used. Chopped strand mat and resin are poured inside the mold, which is rotated while inside an oven. Rotation distributes the reinforced resin uniformly over the inside of the mold. The process is used for pipe, tanks, and similar cylindrical shapes.

In the pultrusion process, strands of roving are coated with resin in a bath and then drawn continuously through a die orifice of the shape of the finished part. The final operation is an oven cure. Channels, I-beams, rails, and ladder rungs are made by pultrusion.

[9.8.]
matched die
molding

Matched die molding is a hot press method, essentially the same as compression molding, the difference being the lower pressures and temperatures of the matched die process. These pressures are in the range of about 100 to 1000 psi. Mat molding, preform molding, and premix molding are the most common methods in use. Metal dies are employed, usually of steel, cast iron, or aluminum.

In the *mat molding* process, a mat or fabric is draped over the lower mold. It can be impregnated with resin either before pressing or in the press cycle. The mold is then closed to form the article. Heat from the heated mold cures the part. The mat molding process is best suited to large flat and shallow shapes of uniform section such as trays, caskets, and printed circuit boards.

The *preform molding* process uses a fiber glass preformed shape made over a perforated metal screen of the size and shape of the finished article. This screen is supported on a rotating turntable. Chopped roving is blown on to the screen as it revolves slowly. A vacuum is drawn through the perforated screen to distribute the fibers uniformly. The fibers are treated with a resin emulsion binder to hold them together. The resulting preform is next heated to cure the binder so that the preform can be handled. This preform is then placed over the bottom die of a matched metal mold, after being coated with a resin mix of the proper quantity. The press is closed to force the resin through the preform. Heat from the mold produces the cure.

In a *premix molding* operation, a bulk molding compound of resin, reinforcement, and filler is molded in a press. This method is employed for smaller articles, especially if heavy sections or complex shapes are involved. The impact and tensile strength of premixed articles is not as great as those given by a preform, since the orientation of the fiber reinforcement is not closely controlled.

Polyester premix materials have a poor performance in filling intricate molds because they are not free-flowing. Higher molding pressures, therefore, are commonly used with premixes.

A *prepreg* (preimpregnated) is used to produce a preform. The prepreg is a sheet of glass fiber preimpregnated with a catalyzed resin which can be cut into patterns. Diallyl phthalate is commonly used in prepregs.

A premix compound is normally used for the injection molding of fiber glass reinforced thermoplastics. Fiber glass is impregnated with the thermoplastic resin and then chopped into pellets. The tensile strength of the material is at least doubled by such reinforcement, rigidity is much greater, and shrinkage, creep, and thermal expansion are reduced. Transparency of the original thermoplastic is, of course, lost by reinforcement. Reinforced thermoplastics are injection-molded, but rarely extruded. Random orientation is a disadvantage of extrusion.

[9.9.]
mold and part design

Thickness of the part may vary from 0.30 to 0.25 in. in preform molding, or as much as 1 in. for premix molding. Thin walls are less subject to such defects as blisters, cracks, and warping. Tolerances of 0.005 in./in. are possible on dimensions, though not across parting lines. Any radius should be as generous as possible, regardless of the method of molding. It is not possible to deposit glass fiber reinforcement in a sharp corner. Draft of at least one degree should be provided for ease of withdrawal of the part from the mold.

Shrinkage of the formed article produced by premix and preform molding is influenced by factors other than cooling of the resin, including the quantity of reinforcing. Variations in glass-to-resin ratio and the orientation of the reinforcement may result in irregular shrinkage and warping. Uniform shrinkage is promoted by a uniform wall thickness. Changes in wall thickness should be produced by tapering from one thickness to the other. Large flat areas must be avoided, because of the probability of warping.

[9.10.]
typical properties

The different methods of molding reinforced articles produce variations in mechanical properties. Some values are given in Table 9.1. Filament winding

Table 9.1

	E-VALUE FLEXURE 10^6 PSI	TENSILE STRENGTH 1000 PSI	ELONGATION PERCENT
Epoxy, filament wound	5–7	80–250	1.6–2.8
Polyester, preform mat	1.3–1.8	10–24	1.0–1.5
Polyester, mat, spray-up	1.0–1.2	9–18	1.0–1.2
Polyester, premix	1.5–2.5	5–10	0.3–0.5
Diallyl phthalate, mineral fill	1.2	4–7	0.0
Polyester, mineral fill	2.2–2.7	3–5.5	0.0
ABS, fiber glass		19	3–4
PS, fiber glass		12	2–3
Acetal, fiber glass		13	2
PP, fiber glass		7.8	2–3
Thermoplastic polyester, fiber glass		19.5	3–4
PC, fiber glass		18	4–6

gives outstanding strength, and the premix process the lowest strength and elongation. The strength of reinforced thermoplastics is roughly double that for the unreinforced resins.

QUESTIONS

1. Are the following properties increased or decreased if a thermoplastic is reinforced with fiber glass? (a) E-value; (b) tensile strength; (c) elongation; (d) coefficient of thermal expansion; (e) creep resistance.

2. What advantages do reinforced plastics offer over metals?

3. What is the difference between a preform, a prepreg, and a premix?

4. What effects will excessive evaporation of styrene produce in a product molded of polyester?

5. What adverse effects are produced by moisture in filler materials?

6. What is the difference between an open and a closed mold process?

7. Will a male or a female mold give least shrinkage? Why?

8. What is a gel coat? A mold release? An internal mold release?

9. What effect does temperature have on cure time of polyesters?

10. What precautions would you take in the design, specification, and manufacture to ensure weather resistance of a glass-reinforced product?

11. What is the difference between matched-die and compression molding?

12. Which of the methods discussed in this chapter would be used to produce the following reinforced products? (a) An auditorium chair; (b) a cafeteria tray; (c) a well casing (pipe); (d) a naval hull; (e) an eavestrough.

10

press molding
of plastics:
compression
molding

[10.1.]
molding of thermosets
and thermoplastics

Long continuous shapes in plastics, such as pipe, are extruded. Bottles and containers are blow-molded. Sheet can be formed to final shape by vacuum-forming. Most plastic products, however, are formed from powder or pellets in a mold held under pressure in a hydraulic press.

The standard method of molding thermosetting plastics is *compression molding*. For thermoplastics the standard molding method is *injection molding*. Different machines are required for the two methods. A thermosetting resin must polymerize during molding, whereas the thermoplastic, already polymerized, merely cools in the mold after being heated to a fluid condition.

Either in compression or injection molding, parts may be made one at a time (one per press cycle) in a single-cavity mold, or many identical parts may be produced in what is called a multicavity mold.

Plastic molds are made from special types of mold steels that can resist the pressures and temperatures of the plastic molding cycle. The temperatures of molding may range to 600° F or higher, and ordinary steels will soften at such elevated temperatures. The die cavity may be machined on a milling machine, spark-eroded on an electric discharge machine, or hobbed. The hobbing process uses a hardened steel master with the shape of the plastic component to be molded. This master is forced into the block of softer die steel at slow speed under the pressure of a large hydraulic press. The cavity produced by hobbing is afterwards ground to exact size and given a polished finish after a hardening operation. Imperfections and scratches in the mold surface will be reproduced in the plastic parts taken from the mold, and if the mold surface is not smooth, the soft plastic will adhere to it. A chrome-plated surface is a low-friction surface, and many molds are chrome-plated to prevent sticking of the hot plastic.

Mold releases, also called release agents, are also used in the molding of plastics. These releases may be waxes, silicones, stearates, or other nonsticking materials. Release agents are blended into the plastic molding compound (internal release agents) or applied to the mold by spray, brush, or other means. Releases must not adversely affect curing nor produce discoloration or other defects.

Often plastic parts must hold close dimensional tolerances of a few thousandths of an inch. This would be necessary in the case of nylon gears for office machinery, for example. For close dimensional control the shrinkage of the plastic as it cools must be allowed for in dimensioning the mold. For example, if a certain dimension on a thermoplastic part must measure 1.000 in. and the thermoplastic shrinks 0.007 in. on cooling from molding temperature to room temperature, then the corresponding dimension of the mold must be 1.007 in.

[10.2.]
product design for press molding

There are three aspects of product design:

1.   Artistic design for eye appeal, which is not discussed in this book.

2. Engineering design, which is the preoccupation of the first half of this book.

3. Design for manufacturing. Here we shall be concerned with design for the molding processes, since these processes are the most important of the manufacturing methods used with plastic products.

When one is designing a molded product, the first and most important consideration is that of how to remove the piece from the mold. In the press methods of compression and injection molding, the part is released from the mold by ejector pins. Therefore, no undercuts can be designed on the interior or exterior of the part unless the part is sufficiently flexible to be sprung out from such undercuts. Undercuts may be managed with the use of a split-cavity mold, but such a mold is expensive. Threaded sections may be used in the plastic part, because threads can be unwound.

A fin or flash of extruded plastic will be produced at the plane where the two halves of the mold meet. This plane is known as the parting line. The position of the parting line must be wisely chosen. It is preferable to place the parting line at an inconspicuous location if possible. Clearly, however, the parting line must be at the largest cross section of the part if the part is to release from the mold. In the case of a flat or cylindrical surface, the parting line should be located at the top or bottom of the surface. If between top and bottom, the flash will be more difficult to remove and will detract from the appearance of the part. Figure 10.1 shows suitable locations of parting lines.

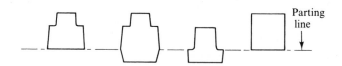

Parting line

FIGURE 10.1 — Locating the parting line on a molding.

Normally a slight draft (taper) is given to the vertical surfaces of the part to facilitate ejection from the mold. A draft angle of 3 deg is sufficient.

When the molded piece requires holes, these are produced by steel pins inserted into the mold. Long slender pins are not practical, because the high molding pressure applied to the plastic will cause the pins to fracture. The depth of the hole should not be more than $3\frac{1}{2}$ times its diameter. A longer hole can be produced by having two pins meet halfway through the hole; this is, of course, possible only for through holes. Long pins must have a draft angle. Side holes can be produced by inserting removable pins in the

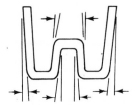

FIGURE 10.2 — A draft angle is required on inside and outside surfaces of a plastic molding.

sides of the mold, but only a restricted length is allowable in a side pin because of the risk of fracture. Angular holes in molded plastic parts are virtually unknown.

Heavy wall thicknesses are not normally designed into plastic parts, whether thermosetting or thermoplastic. Heavy sections require much material and greatly increase the heating and cooling times required in the molding operation. Nevertheless, the wall section must provide adequate strength either by selection of a suitable thickness or by supporting ribs. A defect often associated with thermoplastic ribs (or other changes of section) is the sink mark of Fig. 10.3. Such sink marks can be avoided by making the rib thickness about half the wall thickness, though thicker ribs without sink marks are possible in low-shrinkage and glass-reinforced plastics.

Severe changes in wall thickness should be avoided. See Fig. 10.4.

FIGURE 10.3 — Typical sink mark opposite an increased section thickness.

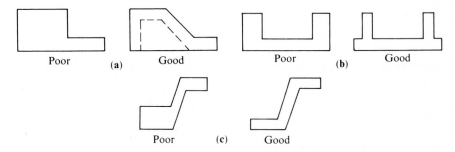

Poor   (a)   Good   Poor   (b)   Good

Poor   (c)   Good

FIGURE 10.4 — Designing for uniform wall thickness.

[10.3.]
compression molding

Compression molding is employed for thermosets, but can also be applied to thermoplastics. Because injection molding is a much faster method, compression molding is not usual for thermoplastics. Thick slabs of thermoplastics may be compression-molded, injection molding being best suited to thin sections. Gears, appliance knobs, handles for kitchenware, gears, toilet seats, and wall switch plates are familiar compression-molded items in thermosets.

In this molding process a predetermined amount of thermosetting compound is placed in a heated mold cavity. The mold is closed by the press while the resin cures and cools. The required temperature and time for curing must be determined by trial, since the thickness of the part section influences the required temperature. Finally, the part is ejected by ejector pins when the mold is opened.

[10.4.]
the compression-molding press

Most compression-molding presses are the upstroke type shown in Fig. 10.5. In this type, a hydraulic ram moves the bed or bolster of the press upward to close the mold. A downstroke press has a fixed lower bolster and a moving upper bolster. The downstroke machines are better suited to the molding of unusually large components, which require a longer stroke.

The strain rods or tie rods hold the upper and lower parts of the press in accurate alignment so that the two parts of the mold do not mismatch. Adjusting collars on the strain rods allow the daylight opening between platens to be adjusted to the requirements of the mold. The daylight opening is the maximum distance between upper and lower platens. This opening must be large enough that the compression molding can be removed from the die.

The two parts of the mold are bolted to the upper and lower platens of the press. These platens have heating channels and are backed with asbestos board insulation. During setting-up the two platens of the press must be checked for parallelism.

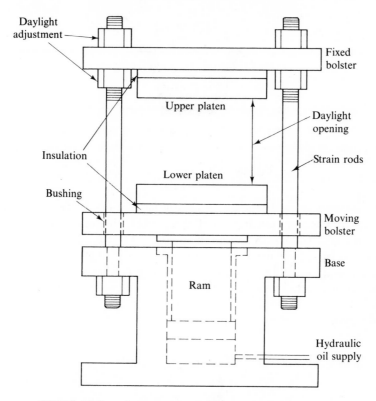

FIGURE 10.5 — A compression-molding press, upstroke type.

The capacity of a compression press is the maximum pressure that it can apply to the mold. Other specifications that influence the capacity and output of the press are the platen area it supplies, and the heating capacity.

[10.5.]
the compression mold

There are three basic types of compression molds.

The *flash* type of mold is used to produce shallow shapes. Excess molding powder is forced out between the top and bottom parts of the mold to produce a horizontal flash (Fig. 10.6). Such flash must be removed from the part after molding. The flash mold does not require as accurate a measuring of the mold charge, since excess material goes to the flash.

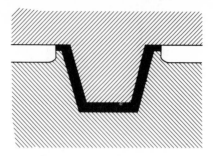

FIGURE 10.6 — Flash type mold.

The *positive* mold produces a vertical flash in the direction of molding pressure. In this mold the upper part of the mold (the force) fits closely into the lower part (the cavity), as seen in Fig. 10.7. The plastic charge must be measured out more precisely for a positive mold. Another disadvantage is that the gas liberated during curing in the mold is trapped, and can produce blisters in the part. The positive mold, however, is suited to plastics of high bulk factor and also to laminated plastics. Clearance between the force and the die cavity is 0.002 to 0.005 in. per side.

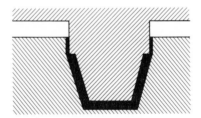

FIGURE 10.7 — Positive mold.

The *semipositive* mold combines the features of both the flash and positive types. External or internal lands (Fig. 10.8) may be used. (The *lands*

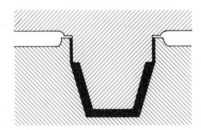

FIGURE 10.8 — Semipositive mold.

are the areas of the faces of the mold that contact each other when the mold is closed.)

The principal parts of a compression mold assembly are shown in Fig. 10.9. The mold assembly is attached to the platens of the press by tee slots or tapped holes in the platen.

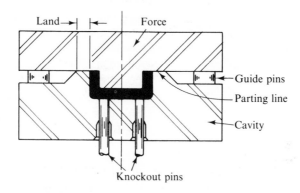

FIGURE 10.9 — A compression mold assembly.

The exact amount of compound required to produce the compression molding sometimes can be calculated from the drawing of the part or a model of the part, the following formulas being used:

$$\text{Volume of part (cu in.)} \times \text{lb/cu in.} = \text{weight (lb)}$$

$$\frac{\text{Specific gravity} \times 62.4}{1728} = \text{lb/cu. in.}$$

Some parts are too complex for such calculations, and for these the amount of compound must be determined by trial. This is done by starting with insufficient compound and gradually increasing the amount until the mold is properly filled. Overcharges cause damage.

The material is fed into the mold either as powder or as a preform, and may be either cold or preheated. Preheating reduces cure time and pressure, and reduces erosion of the mold. During the cure volatile gases must escape, either through mold clearances or vents, or by the mold's being momentarily opened.

Various difficulties develop when the first few parts are pressed in the mold. Molding sticking is one. The material may be overcured or undercured. Overcuring produces a dull or blistered surface, crazing, brittleness, internal cracks, porosity, poor mechanical properties, and color changes. Moisture in the compound may also cause blisters.

Table 10.1
COMPRESSION MOLDING PRESSURES
AND TEMPERATURES

	MOLD PRESSURE, PSI	TEMPERATURE, °F
PF and wood flour	1500–3500	290–350
PF and asbestos	2000–4000	290–350
PF and mineral filler	2000–3000	290–350
PF and glass fiber	2000–6000	290–350
UF and wood flour	4000–8000	275–310
MF and wood flour	2000–8000	280–350
MF and glass fiber	2000–8000	280–350
Polyester	50–300	180–300
Epoxy	100–1000	290–390

Cure time for PF is about 60 seconds for each $\frac{1}{8}$ in. of section thickness. The part is usually ejected from the mold by one or more ejector pins. Depending on the configuration of the part, these may be located in the upper or the lower half of the mold.

The range of compression molding pressures given in the table applies over the projected area of the mold. Clearly, the total force over this area cannot exceed the tonnage capacity of the press. Deep-drawn parts require additional pressure to force the compound to flow up the walls of the cavity. For such deep-drawn parts, 500 to 700 psi is added for each inch of cavity depth beyond the first inch.

EXAMPLE

The part shown in Fig. 10.10 has a projected area of 8 sq in. and a depth of 5 in. What press tonnage is required to produce the part in a two-cavity mold if a PF resin requiring a nominal molding pressure of 5000 psi is used? Allow a 25 percent safety factor in loading the press; that is, the total force must not exceed 75 percent of press capacity, approximately.

Molding pressure on projected area = 8 × 5000 = 40,000 lb
Additional pressure for depth = 4 × 700 = 2,800 lb
Total force   42,800 lb

If a 25 percent safety factor is allowed, about 56,000 lb or 28 tons is the required press capacity.

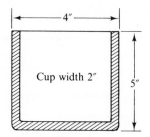

FIGURE 10.10 — A deep-formed cup-shaped part.

Heat is supplied to the mold either by electricity or high-pressure steam. Electric heat is more popular, since it is easy to control and also a cable can follow the movement of the press table without the risk of leakage. With either heat source, heating can be direct or indirect. Indirect heating places the heat source in the press platens, the heat then being conducted through the metal mold to the plastic compound. This method is effective for shallow flat articles. Direct heating means that the heat source is a part of the mold assembly.

In Table 10.1, a range of temperatures is shown. Thin sections should be molded at the upper end of the recommended temperature range, while thick sections are molded at the lower end of the range. The reason is the poor heat conductivity of the plastic resin, and the resulting slow penetration of heat into the part. A lower temperature applied to a thicker section results in less temperature gradient from the exterior to the interior of the part.

[10.6.]
transfer molding

In transfer molding, material is placed in a heated pot, from which it is injected through one or more sprues (channels) into the mold cavity (Fig. 10.11). The sprue becomes waste material and is removed from the part. Pressures in transfer molding lie in the range of 6000 to 12,000 psi.

Transfer molding produces some difference in mechanical properties as compared with compression molding. Because the material must flow through the sprue, fibrous reinforcement becomes oriented in the direction of flow, resulting in higher strength in a direction parallel to flow. Shrinkage parallel and perpendicular to flow direction are not the same.

Transfer molding is preferred to compression molding for parts with wide variations in wall thickness. Such parts are more difficult to cure

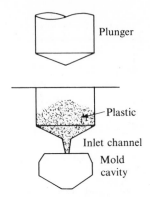

FIGURE 10.11 — Transfer molding.

properly, and require the preheating that is a part of the transfer molding process. If dimensional tolerances are very close, or if flash removal would be difficult in a complex shape, transfer molding is the preferred method. Mold cost, however, is higher for transfer molding.

QUESTIONS

1. What type of thermoplastic part might be compression molded?

2. Explain the difference between compression and transfer molding.

3. What are the advantages of transfer molding?

4. Explain how the transfer molding process makes a part anisotropic.

5. What is meant by daylight opening?

6. For what types of workpiece is a downstroke press suited?

7. Differentiate between a flash and a positive mold.

8. What is the difference between direct and indirect mold heating?

9. Why is flash necessary in compression molding?

10. What are the causes of surface blisters?

11. What is a mold release?

12. Articles made of formaldehyde thermosetting resins are very common, particularly in electrical and kitchen hardware. Find several such products, and attempt to decide the type of mold used to produce the products. Include a wall switch plate for a toggle switch.

11

injection molding

[11.1.]

the injection molding process

Fundamentally the injection molding of thermoplastics, as compared to the compression molding of thermosets, is a simple process. Thermoplastic pellets are heated until fluid, then forced into a mold where the plastic re-solidifies to reproduce the shape of the mold. In practice, however, injection molding is not a simple and easy process, and many details must be suc-cessfully worked out to circumvent difficulties. Time, pressure, and tempera-ture are the basic variables that must be controlled, given the poor thermal conductivity of the thermoplastics. Heating a thermoplastic until fluid is not necessarily easy; it is possible to degrade the surface by heat before the interior of the plastic mass is warm. That the plastic mass will necessarily fill the mold is also an innocent assumption; it may have voids, blisters, sinks, and shorts, and some cavities in a multicavity mold may not receive any plastic at all. Finally, of course, the part must be ejected, and thermoplastics, like thermosets, are adhesive and, given any scope for doing so, will stick.

The plunger type of injection molding machine is shown in Fig. 11.1. The cycle begins with a feed mechanism. This meters out a constant amount of material for each molding cycle. The hydraulic plunger pushes the plastic through the plasticating chamber, which is a heating cylinder, and then through the nozzle into the mold. The torpedo or spreader provides an increased surface-to-mass ratio by reducing the cross section of the passages, thus improving the rate of heat transfer into the plastic.

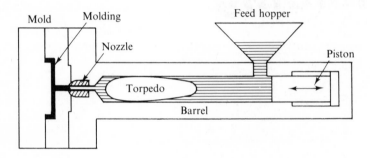

FIGURE 11.1 — Injection molding machine, plunger type.

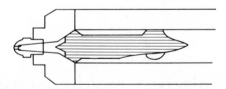

FIGURE 11.2 — The spreader (torpedo).

The fluid plastic is forced into the mold through a nozzle connecting the cylinder and the mold. The area of contact between mold and cylinder is small to minimize the flow of heat from the cylinder to the mold.

Besides a cavity or cavities, the mold must contain passages to connect the cavities to the nozzle. These connecting passages which supply the cavities are the sprue, runners, and gates. The *sprue* is the passage between the nozzle and the runners. It is usually tapered. Each cavity is connected to the sprue by a runner. The gates are very short, restricted passages at the entrance to the cavity. The plastic that freezes in the sprue and runners is removed from the part and reground to be fed again into the molding machine. Figure 11.6 shows these feeding passages.

The principal variables that must be controlled are nine in number. There may be additional controls also, depending on the uses to which the machine is put.

1.   The amount of plastic introduced into the cylinder.

2.   The pressure applied to the plunger.

3.   Plunger speed.

4.   Temperature of the heating cylinder. (Nozzle temperature also may be controlled.)

5.   Temperature of the mold.

6.   Plunger-forward time.

7.   Mold closed time.

8.   Mold clamping force.

9.   Mold open time.

The temperature of the plastic is a critical variable, since viscosity is controlled by temperature. Mold temperature is controlled by a circulating fluid and a means of heating and cooling the fluid. The quality of the molded product is highly dependent on the ability to adjust these control settings. Many of the above variables influence the others; for example, if the temperature of the heating cylinder is increased, then the gate freezes more slowly and a longer plunger-forward time may be needed if sinks and voids are to be prevented.

Immediately after the mold is filled with plastic, the mold pressure is greatly increased, and packing of the mold occurs. The compressibility of the plastic under high pressure allows some flow into the mold after the filling stage. Also, the plastic in the cold mold will at once begin to cool and therefore to occupy less volume and allow more plastic to enter from the gate. The usual practice is to time the ram to return after the gate to the cavity is sealed, in order to prevent any discharge from the mold.

The plastic granules are metered into the process either by volume or by weight. The usual arrangement is a volumetric feed, in which the plastic falls from the hopper into a box, which may be adjusted to the desired volume. The measured volume of material then drops into the heating cylinder in front of the ram.

Volumetric feeding is sufficiently precise for most injection-molding operations. More complex parts, or high-speed operation may require the more precise method of feeding by weight. Slight differences in granule geometry, granule friction, or even static electricity may cause variations in volume.

The plasticating chamber is a steel cylinder heated by electric resistance elements in the outer jacket. The discharge end of the cylinder is fitted with a nozzle that mates with the sprue bushing of the mold. Nozzles may be changed to suit the requirements of the mold. To promote more uniform and more rapid heating, a torpedo or spreader is placed in the plasticating chamber. Heat is transmitted from the cylinder wall to the torpedo by fins.

Some machines are equipped with a preplasticator, which is an auxiliary heating cylinder in which the plastic is melted. The plunger of the preplasticator forces the plastic into the shooting cylinder, which shoots it into the mold. See Fig. 11.3. The preplasticator makes it possible to use lower injection molding pressures, a higher rate of injection, lower cylinder temperature, and a larger shot capacity with increased heating capacity. Some preplasticators use an extruder screw instead of a ram plunger.

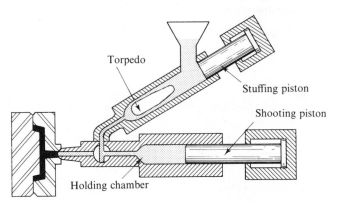

FIGURE 11.3 — Piston type preplasticating machine.

The newer designs of injection molding machines use an extruder screw for melting the plastic. The screw is also used as a plunger to feed the mold. See Fig. 11.4. The extruder screw has a number of important ad-

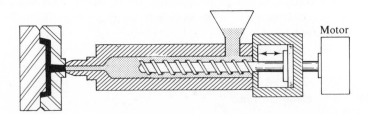

FIGURE 11.4 — Reciprocating screw injection molding machine.

vantages. Mechanical energy from the screw is converted into heat, which is distributed uniformly throughout the plastic as the screw works and mixes it. Heating time and thermal degradation is reduced, and mixing of pigment in the plastic is more uniform.

[11.2.]
machine ratings

Injection molding machines are rated in several ways. The conventional rating is the *shot capacity,* which is the heating capacity of the heating cylinder and the amount displaced by the plunger in one stroke, in ounces of polystyrene per cycle. This may range from 1 ounce to 1000 ounces. The capacity actually used in any molding operation would not exceed 75 to 80 percent of this rating.

Another rating is the *clamping force,* the rated maximum tonnage that closes the mold during the injection stroke. Limitations on clamping pressure may restrict the projected area of the molding. For example, suppose that the clamping force of the machine is 350,000 lb and that polycarbonate is to be molded at 20,000 psi. Then the permissible mold area is limited to a theoretical maximum of 350,000/20,000 or $17\frac{1}{2}$ sq in. However, only a fraction of the pressure produced by the injection cylinder is actually transmitted to the mold cavity, since various pressure losses occur in the heating cylinder, the nozzle, the runners, and the gate. These losses are about one-half of the cylinder pressure.

Thin sections require high injection pressures if they are to be filled. Less viscous plastics such as polyethylene of high melt index flow readily and require less clamping force.

The shot capacity in ounces of polystyrene (or any other plastic) is calculated as follows:

Shot capacity in PS = (volume of molding in cu in.) × (weight of material per cu in.) × (number of impressions)

Allowance must also be made for runners and sprue. The weight of the material in grams/cu in. = 16.4 × specific gravity. The weight of the material in oz/cu in. = 0.578 × specific gravity.

To convert from any plastic to the shot capacity in PS, use the following formula.

Shot capacity in plastic B = shot capacity in PS

$$\times \frac{\text{sp. gr. of B}}{\text{sp. gr. of PS}} \times \frac{\text{bulk factor of PS}}{\text{bulk factor of B}}$$

Table 11.1
PROPERTIES OF PLASTICS FOR INJECTION MOLDING

	BULK FACTOR	SPECIFIC GRAVITY	SPECIFIC HEAT
Acetal	1.8–2.0	1.4	0.35
ABS	1.8–2.0	1.0–1.1	0.35–0.4
Cellulose Acetate	2.4	1.24–1.34	0.3–0.4
PMMA	1.8–2.0	1.17–1.20	0.35
Nylon	2.0–2.1	1.09–1.14	0.4
PC	1.75	1.2	0.30
PE, low density	1.84–2.3	0.91–0.94	0.55
PE, high density	1.72–1.9	0.94–0.965	0.55
PP	1.92–1.96	0.90–0.91	0.46
PS	1.90–2.15	1.04–1.06	0.32
UPVC	2.3	1.35–1.45	0.2–0.28
PVC, flexible	2.3	1.16–1.35	0.3–0.5

The plasticizing rate is the number of pounds of plastic, nominally polystyrene, that the machine can heat to molding temperature in one hour. For any plastic, the plasticizing rate will depend on molding temperature and specific heat. Using PS as the standard and again designating the other plastic as B, we find that

Plasticizing rate using B

$$= \text{plasticizing rate with PS} \times \frac{\text{sp. heat PS}}{\text{sp. heat B}} \times \frac{\text{molding temp PS}}{\text{molding temp B}}$$

Clearly, no machine should be pushed to the limit of its plasticizing rate, but should have some reserve capacity, preferably 20 percent.

Plasticizing rate, lb/hr = weight of molding, lb × number of moldings per hour

The number of moldings that the machine will produce in an hour is somewhat difficult to predict. The molding cycle is chiefly dependent on the cooling period necessary before the mold can be opened. This cooling period is governed by such factors as the weight of the molding, specific heat, temperature of the plastic, thickness of the molding, and its surface area.

EXAMPLE

An injection molding machine has a shot capacity of 32 ounces of polystyrene. What is its shot capacity for polycarbonate?

$$\text{Shot capacity PC} = \text{shot PS} \times \frac{\text{sp. gr. PC}}{\text{sp. gr. PS}} \times \frac{\text{bulk factor PS}}{\text{bulk factor PC}}$$

$$= 32 \text{ oz} \times 1.2/1.04 \times 2.0/1.75 = 42 \text{ oz}$$

[11.3.]
influence of material
properties on the molding

All thermoplastics have a low thermal conductivity. This makes them susceptible to overheating, particularly at a surface in contact with a heat source. If the polymer is exposed to an excessive temperature for a sufficient period of time, the material will degrade. The degradation usually appears as reduced molecular weight and therefore lower viscosity. It is not unknown for a molder who is excessively profit-minded (or simply in order to make any profit) to use a plasticizing temperature that is excessively high, thus improving productivity. The result of such practice is poor mechanical properties and surface defects. If the surface defects are not too serious, the customer may accept the parts, though the poor mechanical properties may not be disclosed until much later. Fortunately, the screw-extruder type of machine greatly reduces the hazard of overheating.

As the plastic cools in the mold, it contracts, so the molding is smaller than the mold in which it was formed. When the mold is being designed from a dimensioned drawing of the part, a shrinkage allowance must be added to the dimensions of the mold. Shrinkage allowances are obtained from the plastics resin producer, usually given as a range of shrinkage values, the exact value depending on the shape of the molding and the injection molding conditions. Rigid PVC has only a slight shrinkage, 0.001 to 0.002 in. per inch, while polyethylene of low or high density has a shrinkage range of 0.015 to 0.030 in. per inch. Polyethylene, PP, and acetal all have very high shrinkage allowances. These materials are crystalline plastics, and in thick sections will cool sufficiently slowly to develop considerable crystallinity.

On the other hand, thin sections of such plastics cool quickly, thus developing little crystallinity and therefore less shrinkage. Tensile strength, density, and modulus of elasticity increase with the amount of crystallinity, while elongation and impact strength are reduced. Thus in these crystalline plastics the shape of the molding has a strong influence on properties.

Shrinkage transverse to the flow direction of the polymer in the mold will be somewhat less than in the flow direction because the long polymer chains become oriented and packed closer together in the flow direction.

Shrinkage is minimized by increased temperature and pressure, by a longer cycle, and by keeping the mold cool.

[11.4.]
the nozzle

Nozzles usually have a spherical nose. This nose must be accurately machined to obtain a satisfactory seal with the mold. The mating spherical recess on the sprue bushing in the mold is usually made $\frac{1}{32}$ in. larger than the radius of the nozzle in order to reduce leakage by ensuring positive seating of the two spherical surfaces at the region of the orifice, and to reduce heat loss from the nozzle to the mold.

The orifice diameter of the nozzle varies with the size of the mold and with the plastic to be molded, but is commonly $\frac{1}{8}$ to $\frac{9}{16}$ in. Normally this diameter is slightly smaller than that of the sprue orifice.

A number of special types of nozzles are in use, including a nozzle with a ball check valve. Reverse-tapered nozzles are used for molding crystalline polymers. These plastics have a sharp melting point, and if too hot they may drool from the nozzle between shots, or if cold will freeze in the nozzle. The reverse-tapered nozzle makes easier the extraction of chilled plastic from the nozzle tip and avoids the necessity of raising nozzle temperature to the point where drooling may occur.

The mixing nozzle of Fig. 11.5 is used in dry-blend color operations. The breaker-plate inserts in the nozzle disrupt the streamline flow pattern to give improved mixing of the pigment.

[11.5.]
runners and gates

From the nozzle, the plastic flows through the sprue bushing in the fixed half of the mold. The *sprue* is cone-shaped for ease of extraction of

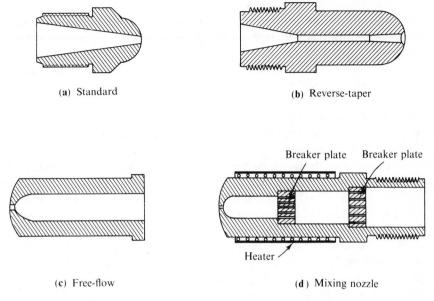

(a) Standard   (b) Reverse-taper

(c) Free-flow   (d) Mixing nozzle

FIGURE 11.5 — Types of nozzles for injection molding.

the solidified plastic. The flow of plastic is distributed to the mold cavities through *runners* (Fig. 11.6). The runners end just short of the cavities and are connected into the cavities by very short channels of small cross section called *gates*. Each gate preferably feeds into the thickest section of the cavity. Vents are required to exhaust air from the cavity.

FIGURE 11.6 — An injection molding of polycarbonate, a component for a small pump. The tapered sprue feeds the cavity through three runners.

The injection mold has a more complex construction than the compression mold, chiefly because of the necessity for sprue and runners. Unlike compression molding, no flash should form during the injection molding process.

The gate is given a smaller cross section than the runner so that the molding can be easily degated (separated from the runners). The positioning and dimensioning of gates is critical, and sometimes the gates must be modified after initial trials with the mold. Feeding into the center of one side of a long narrow molding almost always results in distortion, the molding being distorted concave to the feed. In a multicavity mold, sometimes the cavities closest to the sprue fill first and the farther cavities later in the cycle. This condition can result in sink marks or shorts in the outer cavities. The condition is corrected by increasing the size of some gates so that simultaneous filling of all cavities will result.

Figure 11.7 shows a poor and an acceptable layout for a multicavity mold. In the first part of the figure, the length of runner is excessive. There is a large drop in pressure along the runner; therefore, pressure will be high in the cavities adjacent to the sprue and low in the cavities at the end of the runner. The last cavities may not fill completely. Although the difficulties of the long runner could be adjusted by varying gate sizes, the layout of the second part of the figure is much to be preferred. Besides this problem of

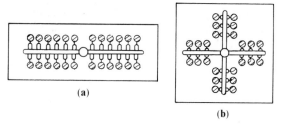

(a)

(b)

FIGURE 11.7 — (a) A poor and (b) a preferred design for a multicavity injection mold. In the linear layout, it will be difficult to fill the cavities at the ends of the long runners.

pressure drop, the mold layout must be symmetrical so that the clamping force is uniformly balanced over the whole area of the mold. Figure 11.8 shows a balanced and an unbalanced layout for a two-cavity mold. The longer runner of the balanced mold does not raise difficulties. Note that the gates feed into the thicker section of the part.

The gate should not shoot directly into an open cavity. If the jet of flowing plastic is not interrupted by a wall or other stoppage, swirls are produced on the surface of the molding. Therefore, the gate should direct the material against some obstruction to break up the flow. Figure 11.9

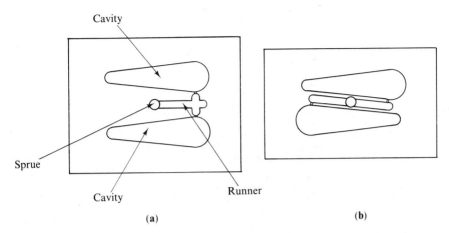

FIGURE 11.8 — The two-cavity design (a) will produce an unbalanced clamping force. The fault is corrected in design (b).

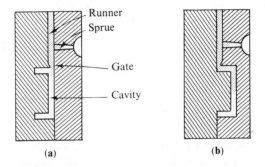

FIGURE 11.9 — In design (a), the incoming plastic shoots into an open cavity, causing jetting. In design (b), the incoming plastic shoots into an obstruction, which is preferable.

shows a layout that allows jetting, and also an improved gating system to break up the flow.

[11.6.]
the mold

The simplest mold is the two-plate mold with a fixed and a moving platen. The three-plate mold has a third plate, which is called the floating or center plate. This center plate carries the gate and some of the runner system.

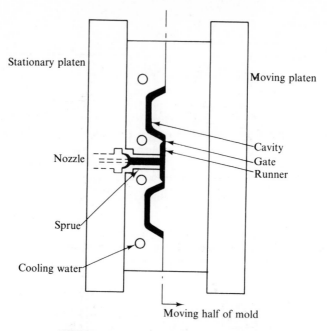

Stationary platen

Moving platen

Nozzle

Cavity
Gate
Runner

Sprue

Cooling water

Moving half of mold

FIGURE 11.10 — Standard two-plate mold.

When the mold is opened, the center plate is separated from the other two plates in order to allow the moldings to be ejected on one side and the sprue and runners on the other. A three-plate mold is shown in Fig. 11.11. Three-plate molds are frequently used in the production of small articles, but in the case of large moldings the center plate becomes too massive. The three-plate mold requires a larger opening stroke; the availability of the necessary stroke must be checked.

The overall dimensions of the mold construction must suit the injection press to be used. The critical dimensions are the closed height of the mold, the opening needed to extract the molding, the width, and the depth. To allow the molding to be extracted, the opening stroke must not be less than twice the height of the molding, that is, the length of the male punch plus the length of the molding plus some working clearance.

Both plates of the mold may be drilled to provide passages for water cooling.

Figure 11.12 shows three methods of feeding a mold. The first method is the preferred one. But if the positioning of the sprue on the exterior of the part will detract from appearance, then the second or the third method must be adopted. In the first method, the molding shrinks on the male punch mounted on the moving platen. In most mold designs the male part of the mold is located on the moving platen, with the molding remaining on the

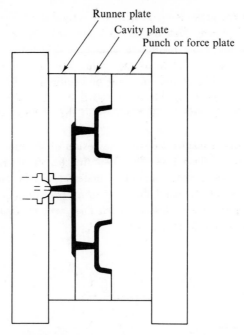

FIGURE 11.11 — Standard three-plate mold.

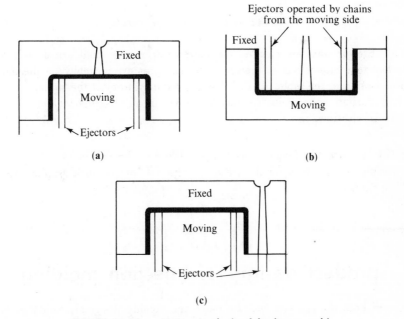

FIGURE 11.12 — Various methods of feeding a mold.

male mold when the press is opened. In the second method the part will be held on the fixed platen when the mold is opened. This method requires a longer sprue. In the third method a side gate is used, and the molding shrinks on to the moving plate of the mold. The flow conditions are less advantageous, and a higher pressure will be required to fill the mold. This may result in increased distortion of the part. Another disadvantage is that the cavity is not symmetrical with the axis of the platens, so that clamping pressure is eccentric.

One of the first considerations in the design of the mold is the location of the parting line or flash line. This split line leaves a small witness line on the molding, since a small amount of material will be able to creep between the two faces of the mold under molding pressure. Consequently, the flash line must be so located that any such flash will be visually acceptable. Some layouts are shown in Fig. 11.13.

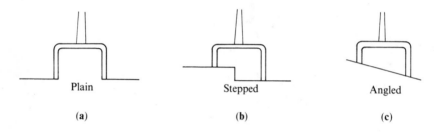

Plain   Stepped   Angled

(a)   (b)   (c)

FIGURE 11.13 — Flash line layouts.

Ejector pins for ejecting the molding from the mold are usually attached to a master bar or ejector plate. The pins are normally on the movable section of the mold. A central pin is required to eject the sprue.

Various types of inserts may be needed in an injection molding. These may be classified as

Blind or through female threads,
   using an insert with a male thread
Male threads

Blind or through holes
Special inserts, such as pins, metal
   stampings, etc.

[11.7.]

product design for injection molding

Unless both the product and the mold are capably designed, moldings will be rejected on grounds either of appearance or properties. Even with good designs, production without rejects or faults is not easily achieved; sometimes the mold must be modified, or the product design modified, or even both. The molding must eject readily from the mold; therefore, undercuts should be avoided and a draft of at least 1 degree should be employed on the walls of the molding.

There should be no regions in the mold where material will solidify before injection of material has been completed. The necessary injection pressure will depend on section thickness, length, and cross section of runners and gates, and also on the flow characteristics of the thermoplastic.

Large thicknesses are difficult to produce by the injection-molding process because of difficulties in filling and cooling the mold, blisters, and sink marks. Products should be hollowed and stiffened with ribs rather than formed in heavy sections. A thick component and its redesign for molding are shown in Fig. 11.14. A uniform thickness results in a more satisfactory part.

Original concept   Redesign

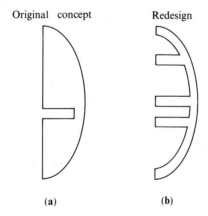

(a)   (b)

FIGURE 11.14 − Redesign of a component for injection molding by adopting ribs and uniform thickness.

For a successful mold to be obtained from the mold maker, he must have more information than a sketch of the part. He needs to know the critical dimensions, tolerances, the special service requirements of the component, how it will be ejected, limitations on gating, and the machine for which the mold is designed.

If the molded surface of the part must be free of blemishes, then internal gating must be used. The mold maker may adopt either subgating or edge gating. For subgating, the gate goes to the component from the parting line, and when the part is ejected, the gate automatically separates. For edge gating, the part, gate, and runner separate as a unit, and the gate and runner must be separated in a second operation.

[11.8.]
colored moldings

Often a product must be produced in several colors. Such changes in color are expensive because it is impossible to clean the previous color from the heating chamber. When a second color is used, some of the former color is absorbed, and moldings of mixed color must be produced until the former color disappears. This interim production of mixed coloring may continue through as many as 50 shots.

Lost production due to color mixing may be reduced by using a clear and colorless grade of the same thermoplastic. The procedure is the following. Suppose that white resin has been used, and a change is to be made to green. All white material is cleaned out of the hopper, which is then loaded with a quantity of clear resin with green resin on top of the clear material. When the clear material reaches the nozzle, it has picked up white pigment, and therefore will be pigmented white. After a time the white color is suppressed and a mix of clear and green material appears. The number of shots of mottled color is greatly reduced by this procedure. The clear material thus serves as a cleaner between one color and the following color.

Pigments and dyes are added as powder to plastic granules and mixed by tumbling in a drum. Each drum must be restricted to its own color. Alternatively, special molding pellets with a high percentage of blended color may be tumbled into the plastic granules. Such colored pellets do not cake like powders and provide a better dispersion of color.

[11.9.]
molding problems

It is not difficult to blend foreign objects or material into plastic granules. The contamination of one polymer by another is a chronic problem because there are so many possibilities for it. Broken bags, spillage, open containers, unclean containers, reuse of containers, and poor housekeeping practices are only a few of the circumstances that lead to mixing of thermoplastic materials. Sprues and runners are reground and used again for molding, and it is easy to mix such regrind material with the wrong resin.

Surface blisters are produced by hot gas in the plastic. Such gas may be produced by solvents, plasticizing agents, moisture, release agents, or even thermal degradation of the thermoplastic. Moisture can be absorbed by pellets in the machine hopper, especially when humidity is high. The cellulosics, acrylics, polycarbonate, nylons, and ABS readily absorb moisture.

Internal voids or bubbles are more common in thicker sections than thin sections. Opaque plastics are preferred because they hide this defect. These voids are caused by trapped gases in the material.

Flash may have a number of possible causes, such as improper closing of the dies, material too hot, excessive pressure, or too heavy a feed. Products with thin walls must be molded at a higher temperature to resist premature chilling in the mold; hence the viscosity is lower and the material can penetrate the parting line.

Sticking of the part to the mold is often corrected by cooling the part of the mold where sticking occurs. Dirt, slight undercuts, or a rough mold surface also result in sticking.

Burn marks, which are charred areas, are caused by hot volatile material such as entrapped air.

Sink marks (Fig. 10.3) are concave depressions, usually at thick sections or on the opposite side to a rib. To prevent sink marks, the gate should be located at the heaviest section of the part. Sometimes the gate must be enlarged to reduce sinks, or pressure may be increased. Often small sink marks cannot be avoided, or may even be acceptable.

Lightweights or shorts are incompletely filled cavities. They may result from inadequate pressure or temperature or too narrow a gate. Other causes may be a clogged nozzle or entrapped air.

A weld line occurs when material flows together from two directions, as around an insert. Weak weld lines result in a defective product, and may be corrected by faster mold filling.

Warping usually results from changes of section. Thinner sections cool fast, and then the thicker sections shrink and pull against the hardened thin sections.

[11.10.]
finishing and regrinding

Flash and fins must be removed from moldings. Definning and deflashing are done in tumbling barrels, the operation also giving a polish to the moldings. Steel stars or balls are used for definning, and sawdust for removing the dust of the definning operation. Pumice, shoe pegs, wax, and other media are used for polishing.

Another difference between compression and injection molding is that in injection molding the waste material, sprues, runners, and rejects, can be regranulated and used a second time. Grinders, also called granulators, are rated by the number of pounds per hour that can be reground.

[11.11.]
injection molding
of thermosets

Thermosets are injection-molded without premature curing by suitable control of heat and pressure and the use of thermosetting resins suited to the process. Plasticating screws are used. Back pressure must be controlled to prevent excessive wear of the screw by the abrasive filler materials common in thermosets. The injection process is basically faster for thermosets than for thermoplastics, especially for thick sections, because the thermosetting plastic cures more rapidly than the thermoplastic can cool in the mold. Other advantages of the thermosets in injection molding are their lower price and their ability to be molded in thick sections.

QUESTIONS

1. What factors rate the capacity of an injection molding machine?

2. Why should the nozzle have only a limited area in contact with the sprue bushing and mold?

3. What is the purpose of the spreader or torpedo?

4. Why are injection machines not operated at maximum capacity?

5. Name some of the thermoplastics that must be heat-dried before molding.

6. Why must granulators be cleaned thoroughly between material changes?

7. Explain the procedure for using a clear resin interposed between two colors in sequence.

8. What advantages are offered by color concentrates over powder pigments?

9. What advantages do thermosets show over thermoplastics in the injection molding process?

10. What is the principal risk when one is injection-molding a thermoset?

11. Why should screw wear be more severe when one is molding thermosets?

12. What are the advantages of the screw plasticator?

13. What are some of the reasons why a successful mold design on paper may not be successful in operation?

14. What characteristics of thermoplastics make them difficult to injection-mold?

15. What is the purpose of (a) the nozzle? (b) the sprue? (c) the runner? (d) the gate?

16. What is a multicavity die?

17. What is the purpose of packing of the mold?

18. If the clamping force is 175,000 lb, and 12,000 psi is required for molding, what maximum projected area of the mold is possible?

19. If the rated shot capacity is 16 oz of PS, what is the shot capacity in (a) rigid PVC? (b) PP?

20. Why will a crystalline thermoplastic that is injection-molded have a greater percentage shrinkage in a thick section than in a thin section?

21. Explain why properties in a crystalline plastic will vary with differences in section thickness.

22. What is the use of the reverse-tapered nozzle?

23. Why is a gate smaller in section than a runner?

24. Explain the subgating and edge gating methods.

25. Why is it preferable to color a thick molding?

26. Why is a molding of very thin section likely to flash?

12

extrusion

Any thermoplastic product required in lengths of uniform cross section is extruded. The range of products produced by extrusion includes pipe, moldings and trim, eavestrough, filaments, sheet and film, wire coating, and blow molding parisons (preforms). The plastic pellets are fed from a hopper into a plasticating screw, which forces the softened material through a die. Cooling and other equipment is required downstream of the die.

[12.1.]
the extruder screw

The most important element of the extrusion machine is the extrusion screw. This delivers material through the die and also heats the thermoplastic by mechanically working it. The length of the screw may be divided into three zones (Fig. 12.1): feed, compression, and metering. The purpose of the feed zone is to pick up the pellets of thermoplastic from the feed hopper and to move them into the main length of the extruder. In this zone the cross section of the screw channel is usually constant. In the compression zone the loosely packed pellets are compacted and softened to produce a continuous stream of molten plastic. In this section the depth of the screw or the pitch may be decreased. Finally the metering zone takes the molten plastic from the compression zone and feeds it at a controlled rate through the die. The metering zone has a smaller channel depth. It acts as a constant-volume metering pump, which explains the term *metering zone.*

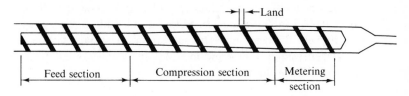

FIGURE 12.1 — An extrusion screw for thermoplastics.

At an open discharge, that is, without a die to create back pressure, the output of an extruder screw would be given by the formula

$$Q = \frac{\pi^2 D^2 N h}{2}$$

where Q = cubic inches per minute
  D = diameter of screw, inches
  N = rpm
  h = depth of screw thread at metering zone
It would seem that the extruder output should be directly proportional to the volume of the last flight of the screw and to the screw rotational speed. This is not quite the case, due to backpressure from the die and other factors.

A general scale factor is applied to the design of extruder screw: If the screw diameter is doubled, then the length must be doubled and also the thread depth. Therefore, screws are characterized by their L/D or length-to-diameter ratio. An L/D ratio of 20 is standard, but some screws are even longer, with $L/D=24$.

Under practical conditions a screw of doubled diameter will have an output increased by eight times. Output is also proportional to revolutions per minute.

While general-purpose screws will extrude a wide range of thermo-plastics, certain plastics impose special requirements on screw design. Crystal-line polymers require a long feed zone, while plasticized vinyls, cellulosics, and some soft plastics require a very short feed zone since they can be compressed as soon as they enter the screw. Nylon melts quickly and re-quires only a short compression zone. Rigid vinyls are degraded by the heating effect of shear stresses in the metering zone, and no metering zone is provided for these plastics, or alternately, a deep channel is provided to reduce shearing effects. Shallower channels give more thorough mixing and generate more frictional heat in the polymer; therefore, the more viscous polymers can use deeper channels.

Most screws have a pitch equal to the diameter and a thread width equal to about one-tenth of the diameter. Clearance between screw and barrel is about 0.0015 in./in. diameter. Since considerable torque is applied to the screw, its root diameter must be substantial to sustain torque loads.

A typical general-purpose screw in a 4½-in. diameter and pitch and 90 in. length would have the following design:

Feed zone	22½	in.
Compression zone	45	in.
Metering zone	22½	in.

The screw material is usually a medium-carbon low alloy steel with hardened flights to resist wear. The screw is hollow so that if necessary it may be heated or cooled by circulating a fluid through it.

The screw is usually driven by an alternating-current motor. Continu-ous variation of screw speed is normally required. Production rate lies within the range of 5 to 15 pounds of product per hour per installed horsepower. A screw of smaller diameter must be operated at a faster speed than one of larger diameter. Speeds of operation range from well below 100 rpm to above 100 rpm.

Since the screw forces plastic through an exit die at high pressure, there is a considerable thrust to be absorbed by a thrust bearing. This thrust bearing requires roller bearings; ball bearings are not sufficiently rugged.

[12.2.]
the barrel

The barrel surrounding the screw must resist the pressure generated by the screw, which may be as high as 6000 psi, or in the case of an extrusion plasticating screw in an injection molding machine, perhaps 20,000 psi. The interior surface of the barrel is hardened to resist water.

Cold water is circulated around the barrel at the feed hopper to prevent the plastic granules from sintering together and thus blocking the feed opening. Usually the length of the barrel must be electrically heated by band heaters around it. While most of the heat supplied to the plastic is derived from the mechanical mixing of the screw, the barrel heaters are necessary for starting up and for additional heat, especially in the extrusion of low-viscosity plastics. Cooling of the barrel is also necessary, especially for film and pipe production. A polymer heated to an excessive temperature could degrade. The cooling system may use water in copper tubes, room air supplied by a fan, or other means. The temperature control system reads the barrel temperature by means of thermocouples.

The extruder pressure is not controlled, but is measured usually by a Bourdon pressure gage tapped into the underside of the barrel. The tube of the Bourdon gage is filled with a grease to keep out molten polymer. Screw speed is measured with a tachometer.

[12.3.]
breaker plate and
screen pack

At the discharge end of the screw the plastic is delivered through a breaker plate and screens to the die. The breaker plate is a thick plate drilled with holes about $\frac{1}{8}$ to $\frac{1}{4}$ in. in diameter. A breaker plate for film or sheet has narrow slots parallel to the die slot instead of holes. The breaker plate streamlines the spiral flow of plastic from the screw discharge and serves also as a support for the screen pack.

The screen pack consists of several layers of stainless wire screen which strain out foreign material and unmelted granular plastic. Foreign matter such as cloth is sometimes found in reground plastic. The heaviest mesh is farthest downstream to support the lighter mesh from being extruded out of the die.

Both breaker plate and screens create back pressure and thus reduce pressure pulsations from the screw.

Polyvinyl chloride readily decomposes at elevated temperatures, and may do so behind a breaker plate. For this and similar plastics, the screens and breaker plate may be omitted and only clean plastic used.

To create high back pressures, a mechanical restriction such as a plug valve may be located at the die head. Such a restriction may be required to improve temperature uniformity or to improve blending and coloring of the plastic through the extruder.

[12.4.]
profile dies

An extrusion die is usually massive. Dies for profiles such as moldings, counter edging, rods, etc. have an orifice of the approximate shape and size of the required contour, the final shape developing outside the die as the material warps and shrinks or expands. Some comparisons of finished profiles with their corresponding die orifices are shown in Fig. 12.2. For complex profiles it is not possible to design a satisfactory die. A first approximation to the required die shape is made, and then by trial on the extruder the shape of the die orifice is modified as required. The size and shape of an extruded profile can also be changed by altering the extrusion speed and the rate of cooling after the plastic leaves the die.

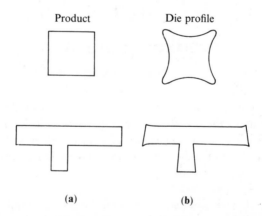

Product　　　　　　Die profile

(a)　　　　　　　　(b)

FIGURE 12.2 — A comparison of two extruded shapes and the die profiles required to produce them.

Critical dimensions for a profile die are the percentage die oversize, the land length, and the angle of entry into the land section. The *land* is the length of constant cross section measured back from the die opening (Fig. 12.3). The land is required to provide back pressure for better mixing. Longer lands are used for very fluid polymers, shorter ones for viscous polymers. Usual land lengths are eight to ten times the thickness (vertical dimension) of the die orifice.

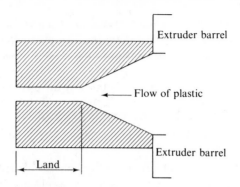

FIGURE 12.3 — The land of an extrusion die.

Usually the die orifice is made oversize. Tension is needed to pull the plastic away from the die, and this tension reduces the cross section of the profile. In addition, the polymer shrinks on cooling. At the same time the plastic tends to swell when it leaves the die and is free of the confining pressure within the die. The crystalline plastics such as nylon and high-density polyethylene require about 20 percent oversize; PS and UPVC require about 10 percent oversize.

Corners and sharp projections in the contour have a large surface area for cooling and therefore shrink first when the contour cools, which tends to make them smaller. This tendency is corrected by enlarging the die at these points. This necessary correction explains the curved shape needed to obtain the square profile of Fig. 12.2.

Profiles are required in smaller quantities than film, sheet, and pipe, and are usually run on smaller extruders. The preferred material for profiles is PVC. This thermoplastic is inexpensive and is not difficult to extrude in complex shapes. However, if the internal passages of the die offer areas where the plastic can stagnate, PVC will degrade rapidly.

[12.5.]
cooling and takeoff
equipment

As the fluid plastic leaves the die, it must be supported by a shaping fixture, usually called a sizing plate, to retain the desired shape through the period when the material cools. For rod and tubing, a metal cooling sleeve or a set of steel or brass sizing plates would be used.

Cooling is provided by a water trough or a water cascade, except for those rigid plastics that can be cooled in air. The air-cooled plastics include PS, PMMA, UPVC, and cellulose acetate; in the case of these materials, water cooling sets up internal stresses and gives a poor surface appearance.

The extrudate is pulled from the die through the cooling and sizing units by takeoff drives. These drives are usually the caterpillar type with two treads. By changing the speed of the takeoff equipment it is possible to control the dimension of the contour; a faster takeoff will produce greater draw-down.

Pipe, tube, rod, and contours are either cut into lengths or coiled.

The following thermoplastics present few problems in extrusion and can be produced in complex contours: the several cellulosics, PS, and UPVC. Polyethylene and PP are more difficult to extrude and are limited to simple profiles. Thermosets can be extruded by using high-pressure ram extruders.

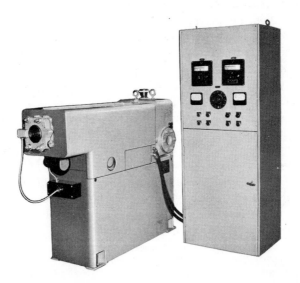

FIGURE 12.4 — A small extruder for producing profiles. The control panel contains temperature and electrical gages and controls.

[12.6.]

pipe

Plastic pipe is ½ in. or more in diameter. Smaller diameters are called tubing.

The pipe die requires a central mandrel or core to support and shape the inside of the pipe wall. This core must be made as a spider with supporting

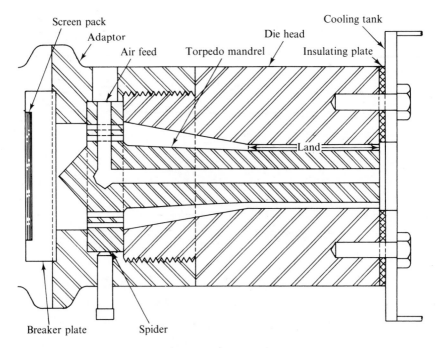

Screen pack
Adaptor
Air feed
Torpedo mandrel
Die head
Cooling tank
Insulating plate
Land
Breaker plate
Spider

FIGURE 12.5 — A pipe die.

arms to hold the core in place concentrically. The plastic flows around these arms and welds together again before leaving the die. The core is water-cooled for PVC pipe in order to avoid degradation, and heated for extruding poly-ethylene.

Several methods of bringing the pipe to size are in use. See Fig. 12.6. A European method is external sizing with a sizing sleeve. The plastic pipe is drawn through a water-cooled metal sleeve and held to the sleeve by air pressure inside the pipe. Alternatively, the pipe may be sized internally by a mandrel, in the extended mandrel method. The shrinkage of the cooling plastic pipe keeps a tight contact with the internal metal mandrel. The mandrel is water-cooled. This method is well suited to polyolefin plastics.

The sizing plate method is now less used than formerly. In this method the pipe takes its size by passing through a series of metal sizing plates. The newest and most used method of sizing is the vacuum trough. The pipe is fed through a long closed water-filled trough. Metal sizing rings in the trough give the pipe its desired diameter. A vacuum is drawn over the water in the trough to reduce the pressure within the trough, and this reduced external pressure expands the pipe against the sizing rings. The inside of the pipe is at atmospheric pressure.

Air pressure within the pipe is a characteristic of some sizing methods.

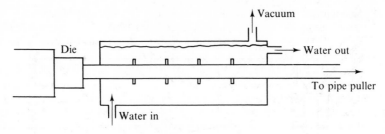

(a) Vacuum trough method

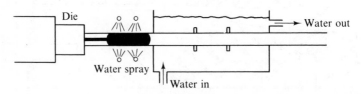

(b) Extended mandrel method

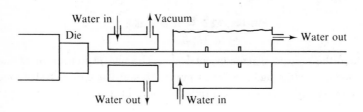

(c) Sizing sleeve method

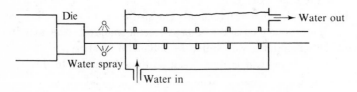

(d) Sizing plate method

FIGURE 12.6 — Methods of sizing extruded pipe.

This air pressure may be retained by the method of Fig. 12.6. A plug is attached to the die with a cord or wire, and the pipe slides over the plug. Other methods are also in use.

[12.7.]
sheet and film

Sheet is produced by extruding the thermoplastic through a long horizontal slit in a sheeting die. The hot sheet is taken through polished metal cooling rolls and then cut to size or coiled. See Fig. 12.7.

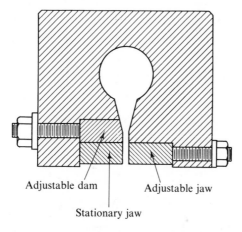

Adjustable dam   Adjustable jaw

Stationary jaw

FIGURE 12.7 — Cross-section of a sheeting die.

Film is considered to be thinner than 0.010 in. Probably the term "sheet" should embrace all flat material extruded from a sheeting die, while "film" should mean flat material blown on film equipment as described below. Narrow strips are slit from wider sheet and film.

Film is produced either as cast film or tubular film. Cast film is, like sheet, extruded through a linear slot die and cooled by contact with a polished metal roll. Most film, however, is blown film.

The production of blown film is shown in Fig. 12.8. The plastic is extruded vertically from an annular die in a thickness usually in the range of 0.015 to 0.025 in. The tube is closed by pinch rolls located high above the extruder; then air is admitted through the center of the die mandrel to blow the tube of plastic into a bubble to thin the tube wall to final thickness, after which the tube is slit to make a flat film. The bubble expands the tube to 1.5–3 times the diameter of the die. A lower extrusion temperature is used for blown film than for cast flat film.

Tubular film is generally tougher and stronger than flat film. The tension of the windup rolls orients the film longitudinally, and air pressure provides orientation in the cross-direction. The transparency of polyolefins also

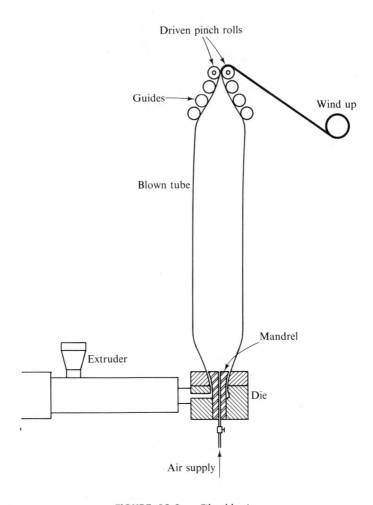

FIGURE 12.8 — Film blowing.

improves with blow ratio (ratio of bubble diameter to die diameter). Best gloss and clarity, however, are given by cast film.

The process for extrusion coating of paper, metal film, or other base stock is illustrated in Fig. 12.9. Adhesion of the plastic coating is promoted by pressure and contact with the chilled metal rolls. To obtain a good bond between the nonadhesive surface of polyethylene and the substrate stock, some oxidation of the PE surface is needed. This is produced by a high melt temperature and adequate distance, and therefore time, between die and contact with the substrate. The substrate material is preheated before coating.

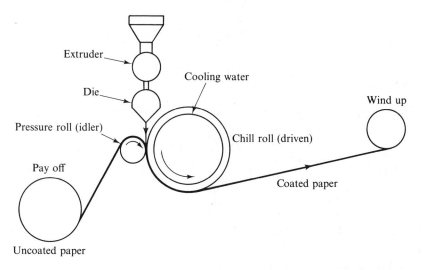

FIGURE 12.9 — Extrusion coating of paper.

[12.8.]

wire coating

Wire and cable coating uses a crosshead die, with the wire travelling at right angles to the flow of the polymer, as shown in Fig. 12.10. The uncoated wire is paid out from a coil and run through a preheater before entering the crosshead die. After passing through the die, the covered wire passes through a cooling trough and spark tester.

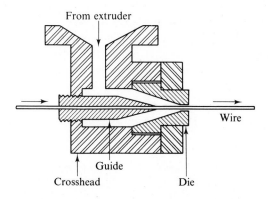

FIGURE 12.10 — Wire coating crosshead.

[12.9.]

extrusion operations

The data in Table 12.1 compare extrusion conditions for different products

Table 12.1

| PRODUCT | HP | BARREL TEMP. | | DIE TEMP. | PRESSURE | MATERIAL |
		REAR	FRONT		(PSI)	TEMP.
pipe	40	150	160	165	1500	165
tubular film	40	150	160	165	1500	165
flat film	40	200	240	250	1000	250
coating	100	250	315	325	1000	320
wire coating	50	220	240	240	3000	240
contours	20	175	200	205	1500	200

in low density polyethylene. All temperatures are Celsius.

It is common to run the extruder to an empty condition when one is shutting down. This ensures that there is no starting up with cold plastic in the machine, a condition that could overload the extruder motor. Some extruders of film prefer to shut down with the screw full of material. This prevents air from entering and oxidizing the plastic. Polyethylene is preferably left in the machine during shutdown.

Because PVC readily decomposes with heat, to ensure that this material is completely removed from the extruder at shutdown, low-melt PE is often used as a purging material. This low-melt plastic can remain in the barrel, since it softens readily on starting up.

When one is starting to extrude, it is preferable to raise barrel temperatures slightly above normal operating temperatures, to be reduced later. The higher temperature ensures that unmelted polymer will not produce excessive torque in the screw.

QUESTIONS

1. Explain the principal function of the feed, compression, and metering zones of an extruder screw.

2. The heat required to soften a plastic can be provided entirely by the frictional heat generated by the working of the screw. However, a plastic

of low viscosity will require additional heat provided from the barrel. Explain why.

3. What are the functions of a breaker plate?

4. In what way might a breaker plate degrade PVC?

5. Why is the contour of an extrusion die usually oversized as compared with the size of the extrusion?

6. Why is the extrusion die for a crystalline plastic considerably over-sized?

7. What is a sizing plate?

8. Why is blown film stronger than flat film?

13

thermoforming

[13.1.]
the thermoforming process

Thermoforming is any process in which a thermoplastic sheet is heated to softening point and forced to conform to the contours of a mold as it cools. Vacuum forming is the best known of the thermoforming methods.

Many plastic components can be formed either by injection molding or by thermoforming. The choice of method is determined by economics.

1. Injection molding uses material in granules, which is the cheapest form of raw plastic. Thermoforming requires more expensive sheet as raw stock.

2. The injection molding die costs thousands of dollars, whereas a thermoforming die, made of aluminum, epoxy, wood, or plaster of paris, costs perhaps a few hundred dollars.

3. If only 1000 units were required, the choice would be thermoforming; a run of perhaps 100,000 or more might be required to write off the cost of an injection molding die.

4. If the plastic components must be produced in a short time, the injection molding method must be rejected because of the long lead time needed to design and produce the die.

5. An injection molding operation will require a day to set up, whereas a thermoforming operation is quickly set up.

6. Thermoplastic sheet can be printed or decorated before forming.

7. Holes or cutouts can not be produced by thermoforming, but are possible in injection molding.

8. Thermoforming is adaptable to the production of very large parts, such as trailer roofs. Deep-drawn parts are also more readily produced by thermoforming.

9. Part thickness cannot always be controlled in thermoforming, because of stretching of the sheet.

10. Many thermoformed parts require a final trimming operation.

11. The thermoforming machine is much less expensive than an injection molding machine, in part because pressures are lower in thermoforming.

The thermoforming process has the following cycles: clamping, heating, forming, cooling, and removal of the sheet. In manually operated machines these operations are timed manually. The basic vacuum-forming machine has a radiant heater, a clamping frame, and a vacuum pump with valve. The mold rests on a vacuum table, which has a vacuum port connecting to the valve. Air pressure may also be provided for forming and for release of the part from the mold.

In the case of a semiautomatic machine, clamping and removal of the sheet are performed manually. The other operations are preset and run automatically with timers and controls. For higher outputs a double stage semiautomatic machine with two vacuum tables may be used. A single heater bank moves back and forth between the two tables, so that one sheet is heated while the other sheet is formed.

Automatic machines are used to produce articles in large volumes, such as packaging containers and housewares. These machines employ rolls of sheet or a stack of precut sheets.

[13.2.]
heating equipment

Infrared radiant electric resistance heaters are the source of heat for the sheet stock. Thick sheet is usually preheated in a hot-air oven to reduce the heating time on the forming machine. The heating time with infrared heaters depends on four factors:

1.   Heater temperature.

2.   Heater power density in watts per square foot.

3.   Distance between heater bank and sheet.

4.   Radiant energy absorption characteristics of the sheet material.

The heat emitted by any heat radiator is proportional to the fourth power of the temperature of the radiating element. Operation at the highest possible heater temperature, therefore, is dsirable, consistent with a long heater life. On the other hand, if the heater temperature is raised, the wavelength of the radiation becomes shorter. Plastic sheets have a lower absorption for shorter wavelengths. However, the increased radiation at higher temperature outweighs the less efficient absorption.

Thicker sheets of thermoplastic, over 0.060 in., must be heated more slowly. Because of poor thermal conductivity, in a thick sheet the surface can be quickly overheated, while the center of the sheet remains unheated for a considerable time after heating begins.

The heating of the sheet accounts for more than 50 percent of the total time of a thermoforming operation.

[13.3.]
the basic forming methods

matched-mold forming —— In this process the sheet is heated, then formed between mating metal male and female molds in a hydraulic or pneumatic press, at 5 to 150 psi. This is a more expensive method, since it requires two molds. It is used chiefly for large sheets with shallow draws. This method gives a more uniform thickness in the part, gives sharper detail, and is a fast method of forming.

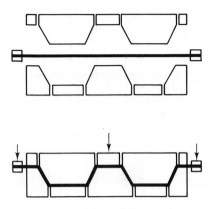

FIGURE 13.1 — Matched-mold forming.

slip forming — Slip forming forms the sheet over a male mold. The process requires both an upper and a lower cyclinder. The loosely clamped sheet is allowed to slip as it wraps on the male mold. Slip forming is selected to avoid excessive thinning of the sheet when parts with deep draws are being formed. Male molds are normally easier to make than female molds, but require more taper on the part because the part will shrink on to the male mold and shrink away from a female mold.

A variant of slip forming is *drape forming*. Here the plastic sheet is draped or forced over the male mold and vacuum is applied to pull the sheet tightly to the mold.

air blowing — Air blowing is a method using compressed air to force the sheet into a female mold. The mold must have vent holes to allow escape of air trapped below the sheet. Air blowing may be combined with a vacuum below the sheet.

vacuum forming — Vacuum forming uses a vacuum to draw the sheet into the female mold. This method may also be employed with male molds. Vacuum forming is the least expensive method, since a mold of only modest strength, such as wood, is needed to resist atmospheric pressure. Straight vacuum forming is limited to draw ratios (the draw depth to the part width) of one-half. Deeper draw ratios result in excessive thinning of the sheet in the drawn areas. Less taper can be used in this method, because the part shrinks away from the female mold.

There are many variants of these basic thermoforming methods. Figure 13.3 shows *plug-assist vacuum forming*, using a male plug. This method

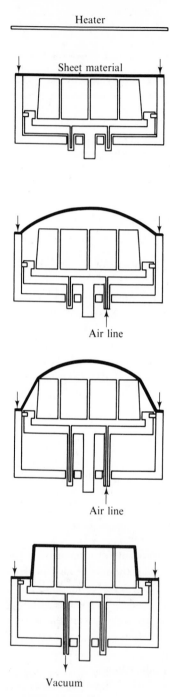

Heater

Sheet material

Air line

Air line

Vacuum

FIGURE 13.2 — Slip forming.

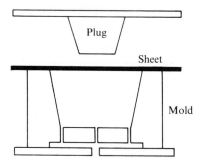

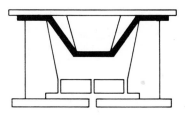

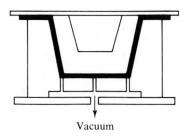

Vacuum

FIGURE 13.3 — Plug-assist vacuum forming.

permits deeper draws with more uniform wall thickness. The plug may or may not resemble the formed article in shape.

In *vacuum snap-back forming* the vacuum is transferred from the female side of the mold to the male side during forming. The plastic sheet is clamped on top of the vacuum box and heated (Fig. 13.4). When the

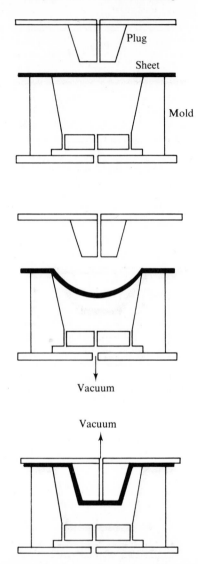

FIGURE 13.4 — Vacuum snap-back forming.

sheet has reached forming temperature, vacuum is applied to the box below the sheet, which is pulled down to a predetermined depth. The male plug next is lowered into the cavity and seals airtight on the vacuum box. Vacuum is shut off underneath the sheet and transferred to the male plug, and the sheet forms against the plug. Wall thickness is more uniform on deep-drawn articles by this method.

Vacuum air-slip forming prestretches the sheet by air pressure between the sheet and the mold. The air bubble is then evacuated and the sheet is drawn to the contour of the mold by vacuum (Fig. 13.5).

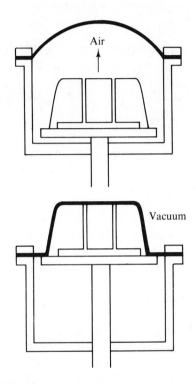

FIGURE 13.5 — Vacuum air-slip forming.

[13.4.]
molds

One of the attractive characteristics of the thermoforming methods is the cheapness and simplicity of the molds. For experimental and short runs, hardwood or plaster molds are suitable. To obtain a better surface on a wood mold, it should be coated with an epoxy resin, then sanded and polished. For longer runs, cast phenolic, cast filled epoxy, and furan molds have been used. Glass fiber reinforcing increases the strength of these mold materials. All these mold materials, wood, plaster, and thermosets, have the disadvantage of low thermal conductivity.

 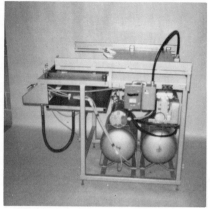

FIGURE 13.6 — A vacuum forming machine equipped with both compressed air and vacuum.

Metal or metal-surfaced molds are used for very long runs. Cast aluminum and magnesium are preferred because of ease of machining and good thermal conductivity.

The finest details are reproduced on the side of the plastic sheet that is in contact with the mold. Location of fine detail, therefore, often determines whether a male or a female mold should be used. If a sheet with a high gloss is to be formed, then the mold surface must be highly polished to protect the plastic surface. A matte surface must be used with polyethylene, however, to allow escape of air, which otherwise is trapped by the high shrinkage of this material. Light sandblasting will give a suitable matte surface.

Shrinkage of the plastic occurs during cooling. The shrinkage will be restricted and therefore less if a male mold is used; the sheet can shrink away from a female mold. The formed part will be more difficult to strip from a male mold. Male molds give a greater thickness at the top of the part; on female molds there is thinning at the bottom of the draw.

Female molds are preferred for multicavity construction. The reason is that male cavities must be spaced farther apart. The distance between male forms should be at least equal to the height of the draw. If they are spaced too close, as in Fig. 13.7, there will be "webbing" (wrinkling) between the forms.

The mold must be drilled with a number of small holes for evacuating or supplying air. Deep corners require the most rapid evacuation, while large flat surfaces require fewer holes. The diameter of the vacuum holes should be 0.010 to 0.025 in. for polyethylene, 0.025 to 0.040 in. for more rigid materials, or slightly larger for thicker sheet. Larger hole sizes will

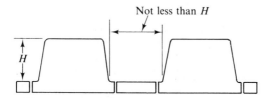

FIGURE 13.7 — Spacing of male thermoforming molds.

produce blemishes. The holes can be counterbored from the back of the die to a larger diameter before the small-diameter drill is used to complete the hole. Wire coated with a release agent is used to make holes in cast or plaster mold, the wires being cast in place and later withdrawn.

Taper on a female mold may be as small as half a degree. A male mold will require a taper of at least 2 degrees on vertical walls, although a smaller taper may be used with air ejection. Corners must be radiused at least equal to the sheet thickness, but for thin sheet no radius should be smaller than 0.06 in. Larger radii are preferred.

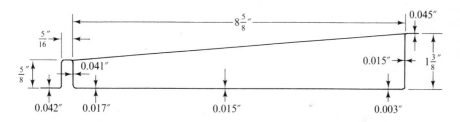

FIGURE 13.8 — Thinning of thermoformed sheet due to the wrong technique of thermoforming. The part was vacuum formed. The polystyrene sheet was 0.040 in. thick.

Stiffness of very thin sheet over large areas is improved by ribs. Shapes with undercuts can be stripped from the mold if the plastic is flexible, for example, polyethylene or plasticized vinyl.

[13.5.]
thermoformed packaging

One of the major applications of thermoforming is the *blister package* of Fig. 13.9. This type of package is produced on roll-fed machines. The transparent film or sheet is vacuum-formed into a suitable shape, such as a rec-

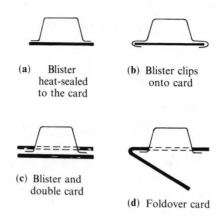

(a)   Blister   (b) Blister clips
       heat-sealed   onto card
       to the card

(c) Blister and
    double card

(d) Foldover card

FIGURE 13.9 — Some types of blister packages.

tangular box, a hemisphere, or the shape of the article to be packaged. The formed "window" is stapled, heat-sealed, or cemented to a printed cardboard in the methods shown in the figure.

In *skin packaging*, there is no mold. The film is thermoformed over the article to be packaged.

The small *portion packs* that supply single portions of jams and jellies on airlines and in restaurants are also used in medicine as unit dose packages. These small containers are also vacuum-formed.

The *shrink pack* is an ingenious variant of thermoforming. Thermoplastic film that has been oriented (stretched) has a memory, causing it to return to its original size when exposed to a flow of hot air. The commodity to be shrink-packed is placed in the film, which is then heat-sealed, and the encased product is transported through a hot-air tunnel which causes the film to shrink tightly around the product. Shrink-packing is chiefly used to make unit loads of cans, newspapers, and cartons, and even to wrap whole pallet loads of goods.

These thermoformed packaging methods offer many advantages:

1.   Improved display of the product.

2.   Sales appeal.

3.   Protection against dust, spoilage, rust, etc.

4.   Easier inventory control.

5.   Easier handling procedures.

6.   Reduced pilferage.

7.   Space for printed instructions or advertising.

[13.6.]
trimming

Perhaps the chief disadvantage of thermoforming is the necessity for separating the thermoformed shape from the remainder of the sheet. This trimming operation can be performed with scissors, band saw, linoleum knife, die punch, guillotine knife, or various die cutting methods. In automated systems the trimming operation is built into the thermoforming machine.

QUESTIONS

1. If the lid to fit a plastic container is to be thermoformed, would you use a male or a female die?

2. Explain the principle of the shrink pack.

3. What is the difference between a skin pack and a blister pack?

4. What mold materials are selected for short thermoforming runs?

5. What advantages do metal thermoforming molds offer?

6. What factors control the heating rate of plastic thermoforming sheet?

14

blow molding and other processes

[14.1.]
blow molding principles

The blow molding process is used to make hollow articles, especially bottles, barrels, and other liquid containers. A tubular preform, called a *parison*, is either injection-molded or extruded and cut to length. The extrusion method is the more usual. The hot parison is placed in a split hollow mold and blown up to conform to the contour of the mold by air pressure in the interior of the parison.

Since only air pressure must be resisted by the mold, aluminum is the usual mold material. Aluminum also offers the advantage of rapid cooling of the article because of its high thermal conductivity. The molds are cored for water cooling. The surface of the mold is sand-blasted to partially hide the parting line, though clear bottles are made in smooth molds. A roughened mold surface also facilitates escape of air between the plastic and the mold. Venting of the air occurs at the parting line, while larger molds also include porous vent plugs.

The blow molding process has four basic operations:

1. Production of the parison.

2. Positioning of the mold halves to entrap the parison.

3. Forming the neck of the container.

4. Injection of air and cooling of the mold.

There are a great many methods of performing these operations and thus a great many different blow molding systems.

FIGURE 14.1 — Blow-molded cosmetic bottle.

The injection blow molding method of producing the parison is preferred to extrusion, where greater accuracy is required at the neck of the opening where a bottle cap may be screwed on. Further, injection molding gives a better surface. The injection-molding method, however, is suited only to small containers below quart size, such as cosmetic jars. Still another method does not use a parison, but two extruded sheets that are sealed together; this method is employed in the production of very large containers and barrels.

[14.2.]
production of the parison

In the injection blow-molding process the parison is injection-molded around a parison stick. The parison stick is hollow, and its far end contains a valve to prevent molten plastic from entering the stick. When the parison has been molded, the mold is opened and the parison on its stick is transferred to the blowing mold. Air under pressure is introduced through the parison stick to expand the parison to the shape of the mold.

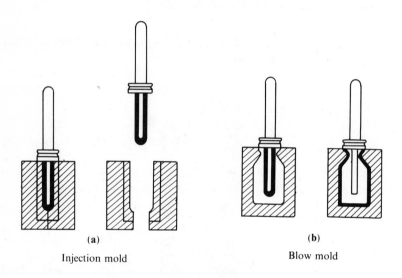

(a)

Injection mold

(b)

Blow mold

FIGURE 14.2 — Injection blow molding.

Two extrusion methods are in use for producing parisons, continuous and discontinuous extrusion. Continuous extrusion is the more usual method, since it applies well to the high-speed production of small or medium-size containers. In discontinuous extrusion the parison is extruded, then the extruder screw ceases rotation while the parison is blown and cooled. The objections to this method are two: The productive capacity of the extruder is not fully used, and there is risk of thermal degradation of the thermoplastic residing in the screw as it waits for the next extrusion cycle.

In the case of a large blow-molded container, the extruder may not be able to produce a parison quickly enough. Even a 55-gallon blow-molded container, which nowadays is not an extremely large size for blow molding, may contain 15 pounds of thermoplastic. Before the parison is completely

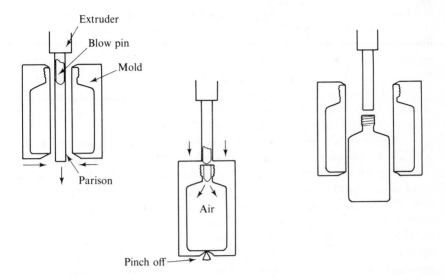

FIGURE 14.3 — Extrusion blow molding.

extruded, the lower end that first came from the die may cool enough that it cannot stretch sufficiently or be closed off to form the bottom of the container. Also, such a weight of plastic would cause the top part of the parison to neck down. For such products the extruder feeds material into a pot or accumulator. The accumulator has a time-controlled hydraulic ram at the top which ejects the parison at high speed through the die while the extruder continues its work of pumping material constantly into the accumulator.

To form the bottom of the container, the bottom of the mold pinches off and seals the end of the parison. Air is usually injected through a mandrel entering the neck of the container at the top of the mold. A completely closed container must be inflated by piercing its wall with a needle orifice supplying air and afterwards sealing the small opening thus made.

[14.3.]
material characteristics
in blow molding

The thermoplastic material enters the mold at a lower temperature than is used in injection molding and is plastic rather than semi-fluid. This simplifies problems of flash and sag. Dimensional control of the containers is a significant problem, especially in those containers that must hold an accurate weight or volume. As the material emerges from the die and is released from die pressure, it swells in diameter and has a wall thickness greater than that of the die annulus. To some degree this swell is compensated for by drawdown due to the weight of the suspended parison, causing elongation and thinning of the walls. In sizing the extrusion die, the designer must take account of these extrusion characteristics of the material.

The die orifice must also be designed from the *blow ratio*. This is the ratio of the maximum outside dimension of the finished part to the maximum outside dimension of the parison after emergence from the die. Low blow ratios of two or three to one reduce the possibility of the walls' blowing out unequally to give variations in wall thickness. However, other factors such as neck diameter usually dictate the dimensions of the parison, and blow ratios may exceed 5 to 1. In all extrusion blow-molded products, a uniform wall thickness is not easily accomplished. Products that do not require best resistance to stress-cracking may be better suited to injection blow-molding, since a uniform wall thickness is more easily accomplished by injection molding of the parison. In an injection-molded parison extra material can be designed into critical areas where thinning will take place.

The extrusion method has the advantage that the extruder can process resin formulations of higher molecular weight than can injection molding machines. Hence, extrusion blow-molding can provide products of better toughness and resistance to stress-cracking.

Two types of fracture are of concern for blow-molded containers: impact fracture and stress-cracking. Good design of the container will improve the strength of the container, as, for instance, by using a generous radius at a corner. Orientation also influences fracture strength. There is considerable circumferential orientation as a result of blow molding. The greater this orientation, the more susceptible will the polymer be to both impact and stress types of cracking. Orientation stress is reduced by increasing parison temperature and mold temperature, and by reducing the blow ratio and the cooling rate. Note that changing the processing conditions influences both types of cracking in the same way. In the case of the crystalline polymers, such as nylon, PE, and PP, increasing the rate of cooling results in less crystallinity and improved resistance to both types of fracture.

A common method of checking impact strength of a blow molding is to fill the container with water and cap it, then to drop the container to a concrete floor from a known height. One container drop from each progressively increasing height is made until an average failure height has been established. All containers must, of course, impact in the same attitude, usually on a bottom edge.

[14.4.]
plastisols

Polyvinyl chloride powder may be plasticized with a large amount of plasticizer to produce viscous solutions called *plastisols*. If the plastisol is heated into the temperature range 330–450° F, it gels to become firm. Such plastisols are used for dip coating of metal products such as tool handles, laboratory ware, and drain baskets for dishes. Plastisols may also be used in the rotomolding process described below.

Organosols are solutions of PVC in a volatile solvent with a small amount of plasticizer. This type of solution gives better penetration into porous materials, and is used to coat textiles, paper, and cardboard. Organosol coatings are commonly found on notebook and book covers. An organosol can be made from a plastisol by adding a suitable volatile solvent that lowers the viscosity and evaporates when heated. The solvent is first flashed off during heating, after which the solution gels.

Plastisol products are made either by dip casting, slush casting, or rotational molding. Rotational molding is discussed in the following section. In dip casting, the article or mold is heated to about 350° F and then dipped into the liquid plastisol, which solidifies and bonds to it. Curing is completed in an oven. In *slush casting*, the liquid plastisol is poured into a heated metal split cavity hollow mold. The mold cavity defines the outside contour of the article to be made. The plastisol solidifies, excess resin is poured out, and curing in an oven follows. The mold is opened to remove the part. Dip casting, therefore, uses a male mold and slush casting a female mold.

[14.5.]
rotational molding

Rotational molding, or rotomolding, is used to make hollow plastic articles such as drums, floats, play balls, containers, tote boxes, picnic coolers, barrels, and furniture. There is no reasonable limit to the size of article that can be produced by this method.

The equipment usually consists of a frame that can be rotated in two planes at right angles to each other, and on which are mounted one or many individual molds (Fig. 14.4). The molds are usually made of aluminum. The required amount of powdered plastic is charged and the molds are

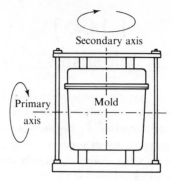

FIGURE 14.4 – The principle of rotational molding.

closed. The charged mold is then rotated about both axes simultaneously as it is heated in an oven. Heat penetrates through the mold walls, causing the powdered material to become semiliquid or to begin to gel. The revolving motion distributes the plastic uniformly over the interior surface of the mold to form a hollow article.

After the plastic has fused into a uniform thickness, the mold is indexed to a cooling station, where it is cooled by water or air. Finally, the mold is indexed to an unloading station where the part is removed. See Fig. 14.5.

There is almost no shape or size that cannot be rotomolded, though minimum wall thickness is limited to 0.030 in. Unlike a thermoformed article, wall thickness is uniform, and internal stresses are absent. Wall thick-

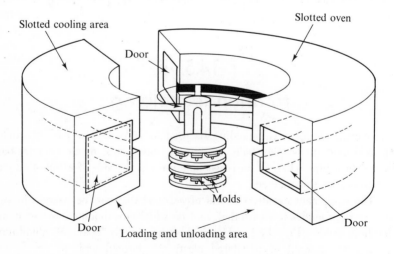

FIGURE 14.5 – Multistation rotomolding system.

ness is controlled by the amount of material charged to the mold. The process can be used with most of the thermoplastics. Sandwich structures are made by rotomolding the outer plastic, then the inner plastic. For example, a solid polyethylene skin may be produced with powder, and then a foamed plastic may be applied behind the solid skin by using a blowing agent.

[14.6.]
powdered plastic

Powdered thermoplastics offer advantages over plastisols in some manufacturing operations. They are easier to handle, are more stable due to absence of solvent loss, and any overspray can be collected and reused.

Metal substrates are powder-coated by three processes: electrostatic spray coating, flame spraying, and fluidized bed coating.

In the *electrostatic spray coating* process, the dry plastic powder is electrostatically charged as it is discharged from a spray gun. The electric charge is negative. The charged powder is attracted to the substrate, which is positively charged. After the part has been electrostatically coated, it is heated in an oven to fuse the coating. Coating thicknesses are in the range of 3 to 30 mils, depending on the service performance required of the part.

In the *flame spray* process, the plastic powder is sprayed through the flame of a gas-fired spray gun that both heats the part and deposits the coating. Flame spraying is used to coat tanks and other articles too large for heating in an oven.

Fluidized bed coating uses the dry powder in a fluidized state mixed with air. The fluidizer tank has a porous bottom through which air or an inert gas passes. This air or gas at low pressure mixes with the powder, holding it in suspension so that it acts like a fluid. The article to be coated

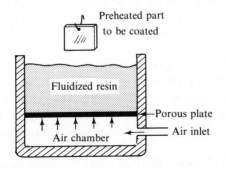

FIGURE 14.6 — Coating parts using a fluidized bed.

is preheated above the fusion temperature of the powder and then is dipped for a few seconds into the fluidized bed. The powder melts and bonds to the heated surface in a uniform thickness. Post-heating may be necessary to obtain a smooth and uniform coating. Coating thicknesses may if necessary exceed 40 mils, usually produced in only one dip. Areas not to be coated can be masked with a fluoroplastic. The process may also be used with thermosets, curing them in a following operation.

[14.7.]
casting of thermosets

Open molds are used wherever possible for casting thermosets to shapes that are easily withdrawn from the cavity. If there are undercuts, then flexible molds of RTV silicone rubber can be used. Metal, silicone, polyurethane, and many other materials have been employed for molds. Though silicone rubber has a nonsticking surface, it is possible for urethanes and epoxies to adhere somewhat to silicone rubber, sometimes for inexplicable reasons. Since silicone rubber does not adhere strongly to the backup material, it is rather easy to strip away a silicone mold lining.

Epoxy adheres strongly to almost all materials. When this material is cast, release agents must be carefully applied to the mold surface. The epoxies otherwise are excellent materials for casting, since their shrinkage is very small. Epoxy tooling for molding metals and plastics is made from casting formulations.

The phenolics also have only a small shrinkage on curing, but are brittle. They can provide beautiful castings. Billiard balls are a good example of phenolic casting and characteristics.

Polyesters are popular for hobby activities in casting, as well as for a wide range of industrial and building products, such as synthetic marble, stone, and brick veneer. Casting polyesters are formulated for low exotherm and shrinkage and minimum tendency to crack. The formulations for fiberglass layup are not well suited to casting because of high exotherm and low viscosity. The high exotherm creates problems such as cracking and distortion if the casting is thick. For heavier castings, less catalyst should be used. Large castings may require refrigeration. Casting polyesters contain a wax that prevents the surface from being tacky. The tackiness of the surface is caused by oxygen from the air, and any method of excluding oxygen will give a hard surface. Mold releases are required with polyesters as with other thermosets.

The pouring of liquid thermoset into the mold must be done in such a way as not to entrap air. Pour into one side of the mold. Vibrating or tapping of the mold may be necessary to remove air bubbles; these bubbles usually rise, and can be pricked.

If articles are to be potted (embedded) in thermoset, either of two methods may be used. The article to be potted may be suspended in position by a thin transparent thread of glass or nylon, and the encapsulating resin poured around it. Or half the compound may be poured, the article laid on the poured material, and the remainder of the material poured over it.

[14.8.]
urethane foams

Rigid polyurethane foams may be poured or sprayed, though the flexible foams are poured. Foaming time must be controlled to suit the forming method. A sprayed foam must begin to foam in a few seconds, while a foam that is to be hand-mixed and poured must not foam until a minute or more has elapsed from the time of initial mixing.

Spraying is the method used to deposit low-density insulating foams on building walls and roofs. Heavier foams such as are used for furniture components (Fig. 7.4) are poured, and the foaming action may be used to pressurize the mold. A foam of nominal weight of 16 lb/cu ft is used to pour the cabinet door of Fig. 7.4, but by using excess material a high pressure is developed in the mold, resulting in a foam density of 22 lb/ cu ft. The higher density is required to provide adequate holding strength for screws and nails. Such a pressurizing procedure will also give a hard skin to the foamed article. The mold, however, must be heavily constructed to resist foaming pressures.

Molded articles may also be produced from froth. The frothed foam is discharged partially expanded, like a shaving cream, and the final expansion in the mold is only about three times.

Foam formulations for pouring may use the one-shot or prepolymer methods. The one-shot method reacts the polyol, isocyanate, catalysts, and blowing agent simultaneously. The prepolymer system reacts the polyol and isocyanate with a mixture of catalyst and blowing agent. There is also a quasi-prepolymer system, wherein the prepolymer is reacted with a combination of polyol and catalyst mixture. Flexible foams use the one-shot method.

Both rigid and flexible polyurethane foam are poured in large free-rising "buns" which may be as long as 100 feet, with a cross section perhaps 3 feet by 6 feet. These large buns are bandsawed into foam cushion slabs or rigid insulating planks. The rigid planks are available in densities of 2, 4, 6, 8, and 10 pounds per cubic foot.

The successful spraying of rigid insulating foam is a special skill, and requires good equipment. Even the best spray guns can clog, and every spray operator has his days when things do not go well. The operator must wear a respirator to keep volatiles out of his lungs. Special primer coats are recommended on wood, which may absorb moisture, or metal, which may be oily, before foam is applied to these surfaces. Successful insulating foam applications are possible only if the operator is familiar with all the circumstances under which he ought *not* to spray, such as a windy day, a dewy surface, rain, low temperature, and others. He must be on guard against wind-carried spray's being deposited on adjacent buildings or parked cars.

[14.9.]
expandable polystyrene molding

The raw material for a polystyrene foam molding is small PS beads containing a blowing agent. These beads are molded in either of two basic methods. The beads may be heated to expand to final low density, then placed in a heated mold and heated to seal the beads together. Alternatively, the beads may be placed in a mold before expansion. The PS foamed cells are closed cells by either method. Densities may be as low as 0.5 lb/cu ft. Perhaps the most familiar article of expandable polystyrene is the foamed coffee cup used in vending machines.

Figure 14.7 shows the process of expansion within a mold. The mold is a split cavity. Steam enters the surrounding steam chamber, then enters the mold cavity through very small holes. The heat of the steam causes the

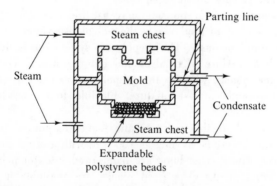

FIGURE 14.7 — Molding with expandable polystyrene beads.

beads to expand to fill the mold. After expansion, the mold is cooled with water and the part is removed. Condensed water and air must be vented, the parting line being used for this purpose.

15

machining
of plastics

[15.1.]
effect of properties
on machining

Wherever possible, the shape required in a plastic article is produced by extrusion, injection molding, or any of the other standard molding methods. Often, however, machining operations are necessary. It may not be possible to mold a hole, so the hole must be drilled in a second operation after molding. Production volume may not be large enough to carry the cost of molds, and the parts may be produced by turning on a lathe, milling, and drilling. Often too, close tolerances can be obtained only by machining after molding.

Suitable machining conditions can be set up only after consideration of the fundamentals of machining practice and the peculiar properties of the thermoplastics and the thermosets. Some plastics are brittle; some are soft and ductile. Drills, milling cutters, and saws will tend to chip and crack such brittle plastics as polystyrene and cast polyester. The thermoplastics have low moduli of elasticity and large elastic recovery, and therefore will deform under the tool forces produced by heavy cuts.

All machining operations generate considerable frictional heat. This basic machining characteristic indicates that machining must be done under conditions that generate least heat, especially for those thermoplastics with low softening temperatures, such as PVC. The poor thermal conductivity of all plastics and rubbers suggests also that machining heat will be confined to the machined surface and not dissipated throughout the material; the temperature rise of this surface can easily be excessive. On the other hand, some thermoplastics have unusually low coefficients of friction, and such plastics should be more easily machinable. Those plastics with the desirable low coefficient of friction are PTFE, PE, PP, nylon, and acetal.

The plastics are softer than the metals, and it is often assumed that tool wear is not significant when these materials are being machined. This clearly cannot be true for a filled plastic. Glass, clay, paper, and other fillers readily wear out tools. Some unfilled plastics are somewhat abrasive. Other tool damage arises from the inability of the plastic to conduct machining heat; the tool must absorb the heat and may be softened by excessive temperature.

The following discussion of machining methods refers to casual machining operations. The setting up of production machining operations for plastic materials may call for entirely different approaches.

[15.2.]

turning

The basic consideration in turning plastics on a lathe is the back rake angle (Fig. 15.1). For casual turning, a back rake of 20 degrees will be found suitable for virtually all thermoplastics. A zero rake angle suits many thermosets. Cuts should not be too shallow or too deep, and speeds of cutting should not be excessive. Shallow cuts develop excessive heat, because the tool rubs as well as cuts. Heavy cuts and high speeds generate too much heat and also cause the plastic to deform under tool forces.

The ductile plastics produce a continuous and flexible chip, which, if not controlled, will tangle around work and toolpost until the cutting area is

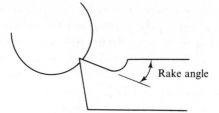

FIGURE 15.1 — Back rake angle on a lathe cutting tool.

hidden by the chip (Fig. 15.2). Such a chip must be steered off the machine to the floor or the pan of the lathe. If the ductility of the plastic allows, a continuous chip should be produced. A discontinuous chip condition is often accompanied by the formation of cracks in the work.

FIGURE 15.2 — Typical tangle of chips when drilling or turning thermoplastics.

[15.3.]
drilling

A drill is the most difficult of all cutting tools to keep cool, since as it drills an increasing length of the tool rubs and generates heat, and as the drill buries itself in the workpiece heat dissipation becomes increasingly difficult.

Drilling heat is the cause of gumming and burning, as well as the development of cracks around the hole. If the chips of hot plastic adhere to the flutes of the drill, then the chips cannot escape from the hole and the temperature rise will be very rapid.

Many thermoplastics exhibit elastic recovery, and this explains why holes drilled in such materials are often smaller than the drill diameter. This elastic recovery may cause the plastic to grip the drill, thus increasing the frictional heat generated between tool and workpiece.

A point angle of 80 degrees (Fig. 15.3) is recommended for drilling most thermoplastics and thermosets. This is a very sharp point compared with those used for drilling metals. The smaller point angle gives less end thrust in the drilling operation.

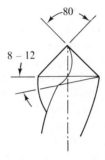

FIGURE 15.3 — Drill angle for drilling plastics.

As in turning, neither too large nor too small a depth of cut per revolution should be used. Small cuts generate excessive heat.

The thermosets, being brittle, are not easy to drill without developing defects such as cracks. The usual small point angle of 80 degrees is recommended to reduce end thrust and cracking in the region of the hole.

Laminated thermosets present special difficulties in drilling. Drilling at right angles to the laminations causes swelling and cracking around the edge of the hole. Drilling parallel to the laminations tends to separate them. These difficulties are reduced by sharpening the drill to 80 degrees. Thermoset laminates will close in the hole when the drill is removed, resulting in an undersize hole.

Molded thermoplastics sometimes have large residual stresses from the molding operation. Such internally stressed parts may crack when drilled. Cracks may be prevented by annealing the part in warm water or air for a time.

The location of hole centers for drilling of metals is done by center-punching. Such center-punching must not be performed on more brittle plastics such as PMMA or PF, since the punch could crack the sheet of plastic. Centers can be marked by scribing two intersecting lines on the plastic or, if the sheet is protected by an adhering paper, on the paper.

[15.4.]
bending and forming

The ductile plastics such as PVC can be sheared without chipping and cracking, though thick sheet may be difficult to shear. Other more brittle plastics must be sawed on circular saws or bandsaws. Generally, when one is sawing the whole area of the sheet should be supported and hand pressure should be maintained on the sheet to prevent chipping and cracking. Considerable heat may be generated in sawing, especially in thick sections. A jet of air is the most convenient method of cooling.

Straight bends in thermoplastics are made by heating along the line of the bend with some type of strip heater and bending while the sheet is hot. Such forming must be done as quickly as possible and followed by immediate cooling. Thicker sheet must be heated from both sides.

When the bend is removed from the bending jig, there may be stress relaxation in the plastic and the bend may tend to open. This tendency may be compensated for by overbending. The degree of overbending must be determined by trial, since it depends on the plastic and its thickness, temperature, and rate of cooling.

16

joining methods

The plastics are remarkably versatile materials, and it is to be expected that a wide range of joining methods is in use for assembling plastic parts. All the methods used in joining metal and ceramic materials are applicable, with suitable modifications, to plastics, including welding, and some special joining methods also are available.

When one is joining by welding or adhesives, proper preparation of the surface is critical for producing a strong joint. If there is a surface film, such as oil or perspiration, a bond to the plastic material is usually not possible.

[16.1.]
adhesives

An adhesive or a cement is a substance that can hold two surfaces together by attachment to both surfaces. Most of the plastics are adhesive, as evidenced by the sticking of molded articles to the mold. Only a few plastics, including PTFE and PE, are not adhesive. Since these do not adhere to other materials under normal circumstances, special methods are required to adhesive-bond them.

There is no universal cement that will bond all types of plastic materials. Five broad classes of adhesives are used for plastics:

1.   Solvent cements.

2.   Bodied cements.

3.   Monomeric cements.

4.   Elastomeric cements.

5.   Thermosetting cements.

Other types of plastic adhesives are available for bonding metals, wood, and construction materials.

[16.2.]
solvent and bodied cements

Solvent cements are used to join a thermoplastic component to another thermoplastic component. The solvent dissolves the surfaces to be bonded. The two surfaces are pressed together, and after evaporation of the solvent, the joint is completed.

Methyl chloride and methylene chloride are often used as solvent adhesives. The resins best suited to solvent-bonding are the amorphous plastics such as ABS, acrylics, cellulosics, PVC, PS, and PC. The thermosets cannot be joined by this method.

If the two surfaces to be joined do not fit well together, such that there are gaps to be filled by the cementing material, then a bodied cement must be used. Since the solvent will evaporate, it cannot fill gaps. A bodied cement is produced by dissolving some of the parent resin in the solvent before applying the solvent to the joint.

A monomeric cement is made from the same monomer as the thermoplastic to be joined. For example, PMMA may be bonded with methyl methacrylate. This monomer is applied as a solvent cement, but may contain a catalyst to cause it to polymerize in the joint. The bond is then produced by polymerization rather than evaporation.

Solvent cements do not provide the strongest and most reliable types of joints. Because the joint is produced by evaporation of the solvent, shrinkage stresses are set up in the joint as the evaporation proceeds. Evaporation may also result in voids within the layer of adhesive.

[16.3.]
elastomeric cements

Movement between the two bonded components, such as may arise from thermal expansion, may break a cemented joint. To ensure that the bond will hold under such conditions, an elastomeric adhesive must be used. The elastomers are capable of sustaining very large elastic strains at low stress levels. This characteristic makes them suitable for joints that must sustain great strains, such as expansion joints in large structures such as buildings.

For extremes of movement, only elastomeric adhesives applied in thick joints will serve. This design principle is best explained by an example. Suppose that an expansion joint must absorb a movement of 0.1 in. perhaps because of thermal expansion. Suppose that the cured adhesive is 0.01 in. thick. Then the adhesive must sustain a strain of 1000 percent (0.1/0.01). Now suppose that the adhesive is applied 0.2 in. thick. In this case the joint must be capable of absorbing a strain of 50 percent (0.1/0.2). Most rubber polymers can easily strain 50 percent in a stress-strain test, but this is too much strain to demand of an elastomer over a period of years of weathering and aging. The example, however, substantiates two design principles for joints:

1. Excessive movement requires the selection of an elastomeric adhesive.

2. The greater the movement to be absorbed, the thicker the adhesive must be.

The elastomeric adhesives may be of natural, synthetic, or reclaimed rubber. Butadiene-styrene rubber is often used. The elastomer may be dissolved in a solvent or emulsified with water. The neoprene adhesives are contact or pressure-sensitive adhesives; that is, they stick immediately on contact. Pressure-sensitive adhesives are also made of SBR, but this elastomer is not recommended for joints that must sustain tensile or shear forces.

[16.4.]
thermosetting adhesives

The thermosets are insoluble in most solvents after they have been cross-linked. These materials must be bonded with thermosetting adhesives, also called reactive adhesives. The thermosetting adhesives are also extensively used to bond metals and construction materials. They produce joints of very great strength, hence their use in the aerospace industry, which may use as much as half a ton of thermosetting adhesives in the assembly of a large aircraft.

The various formaldehyde resins are the most used for adhesive applications. Phenol-formaldehyde has the best weather and water resistance, which explains its use in exterior grades of plywood. If its dark color is not acceptable, melamine or urea formaldehyde may be used, but these do not provide equal resistance to water and weather attack.

The epoxies are more expensive, but are outstanding in applications requiring joint strength, hardness, weathering ability, and bonding to ill-fitting surfaces. They have negligible shrinkage on curing and thus do not introduce shrinkage stresses into the joint. The strength of an epoxy joint is independent of the thickness of the adhesive, so that epoxies can perform a dual service as an adhesive and a void filler.

[16.5.]
characteristics of adhesives

A *substrate* is a material on which a coating, such as an adhesive or a paint, is spread. An *adhesive* is a material that will wet and join two surfaces by attaching to them. A *cement* is the same as an adhesive.

Tack is the property of an adhesive that causes a surface coated with the adhesive to form a bond immediately on contact with another surface. Tack is thus the stickiness of an adhesive.

Peel strength (Fig. 16.1) means the resistance of an adhesive to stripping, and is expressed as pounds per inch width of adhesive joint. Although

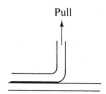

FIGURE 16.1 — Peel of an adhesive.

many adhesives, especially epoxies, have shear strengths exceeding 5000 psi, the peel strength of any adhesive is low. All adhesive joints must be protected against peel stresses. This may be done with rivets or other types of fasteners. Adhesive joints also are relatively weak in tension. The joint design, therefore, must ensure that the adhesive is stressed in shear.

Pressure-sensitive adhesives are frequently made up in tape form on a backing material. Such an adhesive holds after a brief application of light pressure, its dominant characteristic being tack. The adhesive is some type of elastomer compounded with tackifying substances so that the adhesive remains soft, viscous, and permanently tacky. Pressure-sensitive adhesives are unusual in that they do not wet the substrate. They cannot sustain any continuous stress, however small, and have no resistance to peel. Their useful life is limited to perhaps one year.

[16.6.]
hot-gas welding

In the joining of plastics, a distinction is made between welding and sealing. Welding means the joining of plastic components of relatively heavy thickness, whereas sealing is the term usually reserved for the joining of film and foil.

Since thermoplastics can be softened by heat like metals, with suitable modifications the techniques of gas-welding of metals may be applied to thermoplastics. The usual types of gas welding torches employed for metals have flame temperatures that are much too high for welding plastics. The oxyacetylene flame has a temperature of about 6000° F. The low thermal conductivity of the thermoplastics calls for quite low welding temperatures so that heat can penetrate into the plastic from the surface before the surface is degraded.

Metals, being crystalline, have a sharply defined melting point. Thermoplastics usually have a wide range between softening temperature and the temperature at which they degrade. Thermoplastics that have a wide "melting range" are the easiest to weld, though almost all thermoplastics may be welded. Most plastic welding, however, is performed on PVC and PE.

The basic welding method is the application of a heated gas from a gun, usually air or nitrogen, applied to the edges of the sheets to be joined. The filler metal is an extruded rod of the same material as the sheet to be welded. Some degradation of the plastic occurs, so that a plastic welded joint is not quite so strong as the parent plastic. Welding speeds that are too slow will decompose the polymer; speeds that are too fast will result in welds of poor mechanical strength. To maintain weld quality, several fast passes rather than one slow pass may be indicated. Actual welding speeds are slower than those customary in welding metals, about 1 to $2\frac{1}{2}$ in. per minute.

The welding gun (Fig. 16.2) supplies heat to the welding gas either from an electric heating element or a coil heated by a gas flame. The electrically heated guns are lighter in weight and more compact and therefore are preferred, though the gas-heated guns may be necessary for field work where an electrical connection is not available. The guns are supplied with tips with a range of orifice sizes. The welding temperature is varied by adjusting the distance of the gun to the work, or the gas flow, or by a variable transformer. Gas consumption lies in the range of 0.2 to 1 cubic foot per minute of welding time. A reduced gas flow will increase the gas temperature.

Plastic welding rod is available in diameters of $\frac{1}{16}$, $\frac{3}{32}$, $\frac{1}{8}$, $\frac{5}{32}$, and $\frac{3}{16}$ in. A special elliptical rod is sometimes used for a finishing pass on

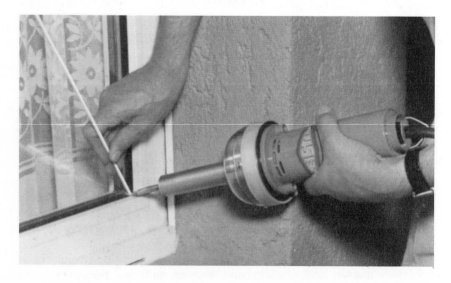

FIGURE 16.2 – Repair welding with a Leister hot air welding gun.

multipass welds. Triangular rod is available for depositing in V-grooves in a single pass. For welding polyethylene the preferred filler rod is a polyethylene blended with 5 percent polyisobutylene. This composition provides better resistance to stress-cracking and a lower softening temperature.

Joint preparation for welding thermoplastics resembles that for metals. The usual types of joints are shown in Fig. 16.3. The single and double vee butt welds use an angle of 50 to 70 degrees with a small root gap. This gap allows the filler metal to penetrate into the root of the weld to give 100 percent penetration. The vee should not be bevelled to a feather edge; such an edge would be decomposed during welding. A nose of about $\frac{1}{32}$ in. is sufficient. The cut edge of the plastic sheet must be clean, but solvents must not be used.

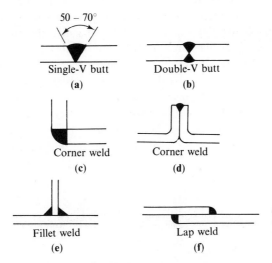

FIGURE 16.3 — Types of plastic welded joints.

The reinforcement (deposited filler metal standing above the surface of the sheet, as shown in Fig. 16.4) should preferably not be removed, since if this is done there is a loss of strength in the joint of about 20 percent. The reinforcement should be removed only if appearance, but not strength, is critical.

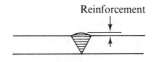

FIGURE 16.4 — Reinforcement of a weld.

[16.7.]
the welding operation

The plastic filler rod is cut to length with a knife. The required length is slightly longer than the length of the welded seam. The cut end is sliced at an angle of about 60 degrees to provide a thin wedge that is easy to heat and convenient for beginning the weld.

Air is the usual welding gas, except for PE and PP, which require nitrogen. The welding torch tip is held at a distance of $\frac{1}{4}$ to $\frac{3}{4}$ in. from the sheet being welded. The sheet and rod are preheated. The rod is held at right angles to the sheet, or 45 degrees in the case of the polyolefins. When the rod is at proper temperature, it will stick to the sheet. As welding continues, a slight pressure is applied to the rod, and a weaving motion from rod to sheet is given to the torch to distribute heat and prevent overheating. See Fig. 16.5. The rod is heated only at the surface and is not melted throughout, as occurs in the welding of metals. Welding defects and weakness commonly occur at the start of the weld. Therefore, in multipass welds different starting points should be used for each pass. The rod angle of 90 degrees is used because lesser angles to the sheet produce cracks. For a fillet weld the rod should bisect the angle between the two sheets.

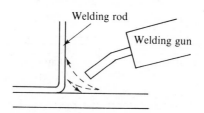

Welding rod

Welding gun

FIGURE 16.5 — Weaving motion of the welding gun.

At the end of the weld, light pressure must be maintained on the deposited rod for a few seconds until it cools; otherwise the end of the rod may be loosened from the seam.

A method related to hot-gas welding is the *plastic flowgun* process. The nozzle of the flowgun melts the surfaces of the two plastic sheets to be welded and at the same time pushes out a hot plastic filler material to the joint. The extruded hot filler material bonds to the softened edges of the joint. The filler material must, of course, be the same thermoplastic as the sheets to be welded.

[16.8.]
heated tool welding

In this process the two surfaces to be joined are first held against a heated metal surface until sufficiently softened. When the part surfaces are at the required temperature for joining, they are quickly brought together under slight pressure in order to bond.

A variety of heated tools may be used, including knives, soldering irons, strip heaters, and hot plates. The surfaces of the heated tool are usually coated with Teflon to prevent the thermoplastic from attaching to it.

When this method is used to join thick plastic sheet, ¼ in. or more, it is preferred practice to shear the surfaces against each other to eliminate entrapped air. A small bead of flash around the joint may be taken as an indication of improved quality.

[16.9.]
spin welding

Spin welding is also referred to as friction welding. The process is best suited to butt-welding of solid or hollow cylindrical plastic bar. See Fig. 16.6. One of the pieces to be joined is rotated as it is in contact with the other stationary part. The frictional heat generated by rotation and pressure melts and fuses together the two surfaces. No bond will be produced toward the center of a solid bar, since relative motion in this region is minimal.

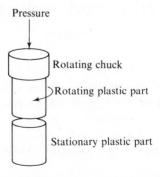

Pressure

Rotating chuck

Rotating plastic part

Stationary plastic part

FIGURE 16.6 — Spin welding.

Small engine lathes and drill presses can be modified for spin welding. Pressure control is most easily obtained from an air cylinder. Figure 16.7 shows some of the types of joints that may be welded if the rpm and pressure are suitably adjusted. The process develops internal stresses, which sometimes must be removed by a heating operation.

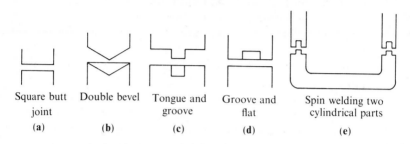

Square butt joint	Double bevel	Tongue and groove	Groove and flat	Spin welding two cylindrical parts
(a)	**(b)**	**(c)**	**(d)**	**(e)**

FIGURE 16.7 — Typical spin-welded joints.

Both rotational speed and pressure must be adjusted to suit the requirements of the thermoplastic that is to be spun. Approximate speeds are given in Fig. 16.8. When sufficient frictional heat has been developed, pressure is increased to squeeze out all voids and the stationary part is then released or the spindle is stopped. The weld cools under pressure. Sawed finishes are suitable, since the rough surface generates heat. Flash may be necessary to develop a strong weld, though often the joint can be designed to trap the flash in an interval recess. The least flash and best quality are obtained in the least welding time, often a second or two.

[16.10.]
hot-wire and induction welding

Two high-frequency electrical methods are used to join thermoplastics: induction welding and dielectric heating. Dielectric heating requires an electric frequency in the range of 30×10^6 cycles per second, and is discussed below. Induction welding may be performed at 10^4 to 10^6 cycles per second. The two joining methods are, however, quite different in principle. Dielectric heating is used to join flat stock; induction welding is best suited to bar stock.

Consider first *hot-wire welding* (Fig. 16.9). A wire of high electrical resistance, such as nichrome wire, is placed between the mating surfaces of

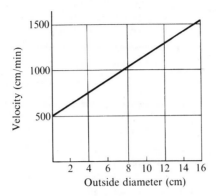

FIGURE 16.8 — Approximate speeds for spin welding.

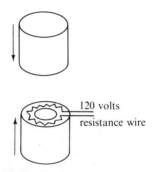

FIGURE 16.9 — Hot-wire welding of thermoplastics.

the two thermoplastic parts. Current flow in the wire generates heat to soften the adjacent plastic surfaces, which can be pressed together to complete the joint. Of course, the heating wire must remain in the joint. For a strong joint, the wire is zigzagged as shown in the figure, or used in multiple loops. The plastic may be grooved to receive the wire.

In induction welding, a circular metal insert, preferably a stamped foil or a wire, is placed between the two mating surfaces. There are, however, no lead wires to take current into the joint. Instead a water-cooled copper coil encircles the joint. High-frequency current passes through this coil. The coil acts as the primary of a transformer, and since the metal insert is the secondary, a current is induced in the insert to soften the plastic. The heating effect is very rapid. Knife blades are attached to thermoplastic handles by induction heating.

Considerable pressure may be necessary to ensure a strong induction weld. Heating cycle time is usually a few seconds. The greater the power

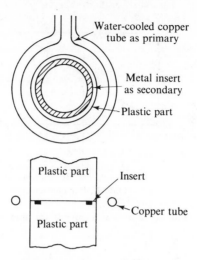

FIGURE 16.10 — Induction welding.

input, the shorter the cycle time, although excessive power input may degrade the plastic adjacent to the heating insert. The insert must be in contact with the faces of the two plastic components over its whole area.

[16.11.]
ultrasonic assembly
of thermoplastics

Ultrasonics is the technology of high-frequency acoustic waves. Its applications use sonic power in the frequency range of 20,000 cycles per second or higher. Few adults can hear sound at frequencies exceeding 12,000 cycles, so ultrasonic vibrations are inaudible. Ultrasonic methods are used for three assembly processes with thermoplastics:

1. Plastic welding.
2. Inserting metal parts into plastics.
3. Staking metal to plastic.

The technique of ultrasonic joining is basically a thermal bonding method using mechanical vibration at ultrasonic frequencies to produce frictional heat between the mating parts to join them. The vibrating tool, usually called a horn, is either wedge-shaped or cone-shaped, often made of titanium, with the small end pressed against the plastic parts to be welded. The tip of the horn may be contoured to match the shape of the part to be welded. Localized heat is produced by vibratory friction at the mating surfaces. No surface preparation is needed before joining. However, a few thermoplastics cannot be welded by ultrasonics, chiefly the vinyls, and the cellulosics are difficult to weld by this method.

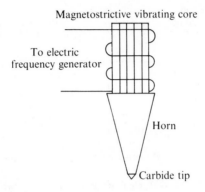

FIGURE 16.11 — Principle of the ultrasonic welder.

Suitable joint designs are shown in Fig. 16.12. The figure shows a butt joint before and after welding. Note the use of the triangular-shaped *energy director*. This localizes the generation of heat so that a restricted volume of the plastic is softened. The small volume of fluid plastic then flows out uniformly over the surface of the joint to produce the bond.

Welding time must be determined by trial. Those plastics with high melting temperatures will require longer vibration times than, for example, PS or PMMA. About a half-second of "hold" time under pressure is needed to freeze the joint.

Ultrasonic staking is used for joining thermoplastics to metal. A plastic stud is mushroomed over the metal by an ultrasonic staking horn that melts and reforms the plastic (Fig. 16.13). Alternatively, the same kind of staked joint may be made with a heated staking tool, though this method is much slower than ultrasonic staking.

Metal inserts can be inserted into thermoplastics by ultrasonics. A hole that is slightly smaller in diameter than the insert guides the insert as it enters under the pressure and vibration of the horn. Inserts designed for this application have knurls or undercuts into which or around which the plastic flows in order to lock the insert in place.

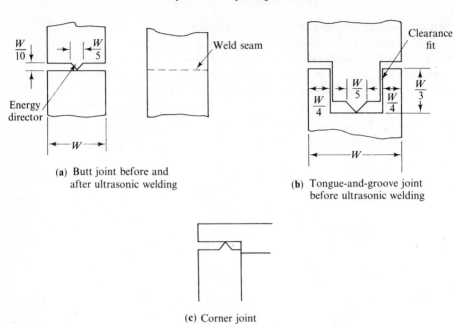

(a) Butt joint before and
after ultrasonic welding

(b) Tongue-and-groove joint
before ultrasonic welding

(c) Corner joint

FIGURE 16.12 — Typical joints for ultrasonic welding.

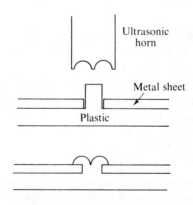

FIGURE 16.13 — Ultrasonic staking.

[16.12.]
heat sealing

Various heat sealing methods are employed for the joining of thin thermoplastic films. These methods are especially important in the packaging in-

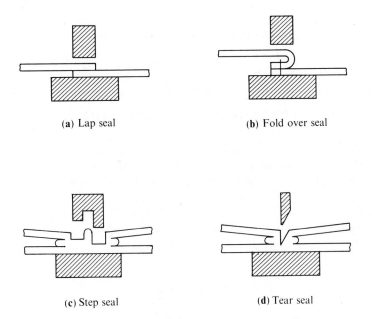

(a) Lap seal   (b) Fold over seal

(c) Step seal   (d) Tear seal

FIGURE 16.14 — Typical seals for heat-sealing processes.

dustry. Some of the more important types of seal produced by these heat sealing processes are shown in Fig. 16.14.

The *lap seal* is used to join two wide sheets of film in a bond where highest strength is required but at some sacrifice in appearance. The *step seal* gives a decorative three-dimensional effect. The figure shows a simple three-step die; more decorative effects can result from more elaborate dies with as many as five steps.

The *tear seal* is familiar in polyethylene packaging at supermarkets. The purpose of this design is to permit easy tearing apart of the film at the reduced section. A *peelable seal* is a weak seal that permits the joint to be torn open without tearing of the film.

The *surfaced* or *serrated seal* is another decorative seal that embosses the film in a regular and attractive pattern. The die is engraved with small cavities that coin the film.

Heat sealing may be done by any of the following processes:

1.   Thermal heat sealing, using a heat tool or die. The reason for using the redundant phrase "thermal heat" is simply to differentiate the method from other heat sealing methods. There is no other kind of heat than "thermal heat."

2.   Thermal impulse heat sealing.

3.   Dielectric sealing, using high-frequency electricity.

[16.13.]
thermal heat sealing

In thermal heat sealing the sealing tool or bar is held at a constant temperature. The basic arrangement is shown in Fig. 16.15. All the heat is produced by the heated tool on one side of the joint and flows through one of the layers of the film to the interface where the joint is made. Since thermoplastics are poor heat conductors, the method is practical only in those cases where one of the films is very thin. For thick material an excessive tool temperature would be required to obtain a joint at the interface.

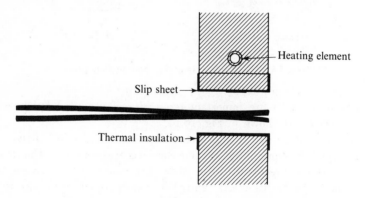

Heating element

Slip sheet →

Thermal insulation →

FIGURE 16.15 — Principle of thermal heat sealing.

A heat-stable resilient material may be used for the bottom die to compensate for variations in film thickness. In other designs both the top and the bottom tool may be heated.

Figure 16.15 shows a slip sheet on the contact surface of the heated bar. The slip sheet is needed when those plastics are being sealed that will stick to the bar at elevated temperatures. Teflon (PTFE), silicone rubber, and silicone grease are all used as antistick agents.

One type of continuous heat sealer is illustrated in Fig. 16.16. The sheet is drawn past preheaters and sealed by pressure applied by small guide wheels. Polyethylene may be continuously sealed at speeds as high as 100 fpm.

Thermal impulse sealing differs from thermal heat sealing in that the heated dies are not held at a constant temperature, but are heated intermit-

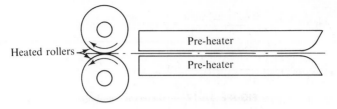

FIGURE 16.16 — Principle of the continuous heat sealer.

tently. The die temperature is increased during the heating cycle and then is dropped to allow the joint to cool under pressure. The die is designed, therefore, for a very low heat storage capacity. The impulse sealing method is best suited for film of a thickness less than 0.010 in. Because the method permits a post-cooling operation, appearance and strength of the joint are improved.

The rapid cooling rate of impulse sealing results in a tough joint in crystalline polymers. These polymers cool quickly, resulting in minimal crystallinity. Impulse sealing, however, is equally useful for amorphous polymers.

[16.14.]
dielectric sealing

The thermal sealing methods supply heat at the surface to flow into the area of the joint. Dielectric heating heats the interior of the thermoplastic instead of the surface. Post-cooling is possible, since the dies operate below the melting temperature of the film. However, certain electrical characteristics are necessary in the polymer material, and not all thermoplastics can be joined by dielectric sealing. Polyethylene, PP, PS, PC, and PTFE do not have the electric properties required for this sealing method.

The thermoplastic to be sealed is placed as a dielectric between two electrodes or sealing bars (a dielectric is an electrical insulator). The electrodes transmit a high-frequency current to the thermoplastic and at the same time exert necessary sealing pressure.

The electrical theory of dielectric heating cannot suitably be explained here; there are many textbook sources for this elementary topic in electrical theory. The basic arrangement is given in Fig. 16.17, where the plastic is shown as a dielectric layer between two flat electrodes and connected across a high frequency power source. The equivalent electric circuit is shown in Fig. 16.18. The capacitance C (capacity for storing electrons) and the re-

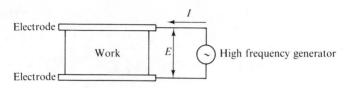

FIGURE 16.17 — Dielectric heating.

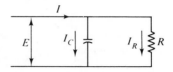

FIGURE 16.18 — Equivalent circuit for dielectric heating.

sistance R of the plastic are assumed to be in parallel. The power developed in the plastic as heat is

$$P = EI_R = E^2/R$$

where I_R is the current through R.

The dielectric heating effect is actually proportional to the frequency, the square of the voltage, the dielectric constant of the material, and the power factor. Hence high frequencies and high voltages are employed. The dielectric constant of the material is defined as the ratio of two capacitances: the capacitance of a parallel-plate capacitor with the given material between the plates as dielectric divided by the capacitance of an identical capacitor with a vacuum for dielectric. Therefore, the electrical resistance and the dielectric constant of a thermoplastic determine its suitability for dielectric heating.

In addition to a suitable dielectric constant and loss factor (loss factor will not be explained here), the plastic must have a sufficiently high dielectric strength to prevent an electric discharge through the plastic under the stress of the high voltage used. Thick material can be heated more rapidly by dielectric heating than thin material, because the heat generated in thin material is readily carried away by the conductive electrodes. The electrodes are commonly made of brass strip up to $\frac{1}{8}$ in. thick backed with brass, aluminum, or steel. The edges of the electrodes are radiused to avoid a concentration of electric energy that would produce arcing and burning. However, tear seals are made by sharpening one side of the electrode, thus weakening the film at this edge to facilitate tearing.

The Federal Communications Commission allots a frequency of 27.18 megacycles to dielectric heating, and most machines use this frequency. Other frequencies may be used, provided that the equipment or room is shielded so that it does not radiate as a radio transmitter.

Dielectric heating is especially useful for joining heat-sensitive materials because the heat is generated rapidly and uniformly throughout the plastic without high temperatures at the surface. Actually, temperature is maximum at the interface between the two layers of film.

[16.15.]
fasteners

A variety of standard and special fasteners are employed in assembling plastic components.

The familiar *rivet* is a standard and inexpensive fastener for plastics. Solid, semitubular, tubular, split, and other types of rivet are all employed for such assembly work, either in manual methods or using bench riveting equipment (Fig. 16.19). The tubular rivet shown in the figure can punch its own hole in soft plastics. The rivet hole should not be closer to the edge of the plastic than a distance of three times the rivet shank diameter. The clinch allowance (Fig. 16.20) should be six or seven tenths of the rivet shank diameter. An inadequate clinch allowance does not provide sufficient upset, while excessive allowance will cause the rivet to buckle.

Blind rivets are used in those circumstances where there is no access to the reverse side of the joint. The blind rivet consists of a hollow body and

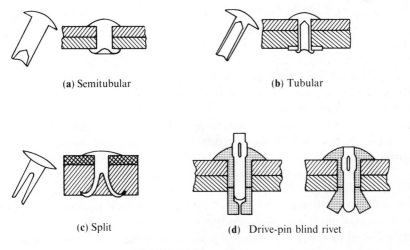

(a) Semitubular (b) Tubular

(c) Split (d) Drive-pin blind rivet

FIGURE 16.19 — Rivet types.

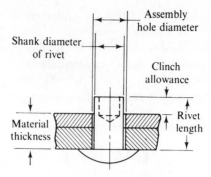

FIGURE 16.20 — Significant dimensions of a rivet.

a solid pin. The solid pin is driven or pulled through the hollow rivet shank to flare the shank on the blind side of the joint.

Self-tapping screws are standard for assembling. There are two types: thread-forming and thread-cutting; both make their own mating threads in the plastic. The thread-forming type is employed with thermoplastics. Since thermosets are too brittle to be deformed by a thread-forming self-tapping screw, the thread-cutting type must be used for these materials. Both types of self-tapping screws are illustrated in Fig. 16.21.

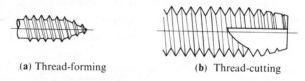

(a) Thread-forming   (b) Thread-cutting

FIGURE 16.21 — Self-tapping screws.

The pilot hole size for a self-tapping screw must be such that the proper torque is developed in the screw. In softer materials a smaller pilot hole is used than for harder materials.

Figure 16.22 shows an assembly that uses a thread-forming screw. Note that the pilot hole is chamfered to prevent swelling of the thermoplastic around the hole.

Bolts and nuts are available in most of the hard and strong metals and alloys, including aluminum, low-carbon steel, stainless steel, brass, and steel plated with cadmium or chromium. Nylon and polycarbonate nuts and bolts also are available. *Cap screws* are quality bolts machined all over to closely controlled dimensions, and available in both hexagon and socket types of head. Both fine thread and coarse thread systems are used, the fine thread

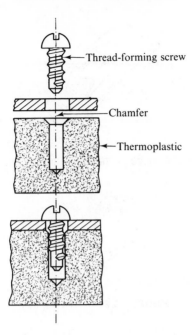

FIGURE 16.22 — Assembly method for thread-forming self-tapping screws.

system being best suited to those circumstances where the nut must not be allowed to work loose.

The term "screw" usually means a small bolt for use in a threaded hole without a nut. A machine screw is an example of a screw, though machine screw nuts are often used. These small screws are slotted for a screwdriver in both round head and flat head styles. Machine screws are designated by number size: 0, 2, 4, 6, 8, and 10, where 0 size is very small and 10 is the largest size.

For protection against moist atmospheres, steel nuts, bolts, and screws are cadmium-plated. Cadmium plating does not protect against severe weather conditions such as rain and snow, and for such conditions aluminum, brass, stainless, or chrome-plated steel must be used.

For lighter-duty applications, single-thread engaging nuts, sometimes called *speed nuts*, reduce assembly time and cost. An example of this type of fastener is the Tinnerman flat speed nut of Fig. 16.23. It requires little torque to be tightened, only enough to produce locking. The screw thread is locked as the arched prongs are compressed. The speed nut is vibration-proof.

The *speed clip* is similar to the speed nut. It is snapped over a molded boss in the plastic part, locking the assembly together securely.

If a nut must be positioned in an inaccessible location, then an *anchor nut* must be fastened in place before assembly (Fig. 16.24). The lug of the

FIGURE 16.23 — Tinnerman flat speed nut.

FIGURE 16.24 — Anchor nut.

anchor nut is often riveted or spot-welded. A nut may be anchored to a thin sheet by the method of Fig. 16.25. This combination is called a *caged nut*. The cage is a boxlike container that retains the nut, restrains it against rotation, and anchors it to the sheet. The *clinch nut* (Fig. 16.26) serves the same purpose, the nut being clinched into the plastic or metal sheet.

FIGURE 16.25 — Caged nut.

FIGURE 16.26 — Clinch nut.

Various vibration-proof locking fasteners, shown in Fig. 16.27, resist rotation by gripping the mating thread. Two general types are available:

Prevailing torque
Free-spinning

The prevailing torque locknut spins freely for a few turns and then must be torqued to seating position. Maximum holding power is obtained as soon as the threads engage the locking feature. A free-spinning locknut is free to spin on the bolt until seated. After complete engagement, additional tightening locks the fastener.

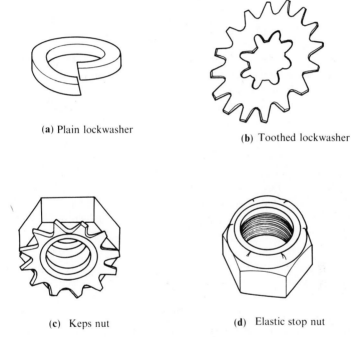

(a) Plain lockwasher

(b) Toothed lockwasher

(c) Keps nut

(d) Elastic stop nut

FIGURE 16.27 — Locking fasteners.

A familiar example of a free-spinning locknut is the common lockwasher of Fig. 16.27. A modification is the toothed lockwasher, which may have teeth on either the inside or the outside diameter. The teeth are twisted out of the plane of the washer face so that sharp edges are presented to both the workpiece and the bearing face of the nut. The Keps nut of Fig. 16.27 uses the same principle; a toothed washer is joined to the nut. The elastic stop nut has a nonmetallic insert such as nylon at the end of the threads that is slightly smaller in inside diameter than the outside diameter of the threaded bolt. When the threads of the bolt enter this soft insert, they impress a mating thread into it. Still another type of free-spinning fastener uses a bolt thread coated with epoxy cement and a nut containing a hardener for the epoxy. When the nut is run on to the bolt, the hardener cures the epoxy and a cemented joint is produced in the threads.

Most prevailing torque fasteners use a deformed thread on one member of the fastener or a soft insert in one of the threaded components, the soft insert being deformed to produce an interference fit.

[16.16.]

inserts

Inserts are a special type of nut serving the function of a tapped hole (Fig. 16.28). The external surface of the insert must be shaped or deformed so that it is locked positively in the hole. A self-tapping external surface is often employed. Other types of inserts may be placed in position ultrasonically, a pilot hole slightly smaller in diameter than the insert being used.

The wire type of insert shown in Fig. 16.28 uses a coil of wire with a diamond-shaped cross section. This shape acts both as an internal and an external thread to hold the screw and to lock the insert into the plastic.

An expansion insert is illustrated in Fig. 16.28. The tapered and knurled bottom end of the insert is expanded by the metal spreader as the spreader is forced down the slots of the insert. Another type of expansion insert is shown in Fig. 16.28, with a screw to expand the insert.

QUESTIONS

1. Why should a thermosetting adhesive be superior in virtually all properties to a thermoplastic adhesive?

2. What are the disadvantages of a solvent type of adhesive?

3. What are the outstanding advantages of the epoxies as adhesives?

4. What is a pressure-sensitive adhesive?

5. Define: (a) tack; (b) peel strength; (c) substrate.

6. Why should an adhesive joint be stressed in shear?

7. What is the difference between a solvent cement and a bodied cement?

8. Why are solvent cements unsuited to the bonding of thermosets?

9. What cement or cements do you recommend for the case of a poor fit between the mating surfaces?

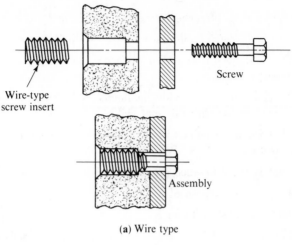

(a) Wire type

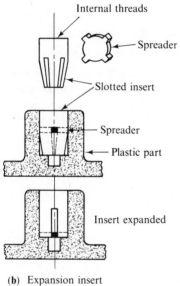

(b) Expansion insert

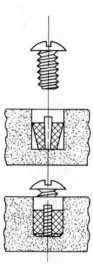

(c) Screw expansion insert

FIGURE 16.28 — Inserts.

10. Why are elastomeric materials preferred for expansion joints?

11. What is the purpose of the slip sheet in heat sealing?

12. Explain the difference between thermal heat sealing and thermal impulse sealing.

13. Why does impulse sealing produce a stronger and tougher joint than thermal heat sealing?

14. Explain the difference between dielectric and induction welding.

15. Why is dielectric sealing a preferred method for sealing thick sheet?

16. Why is dielectric sealing a preferred method for sealing heat-sensitive plastics?

17. Cellulose nitrate is highly flammable. What joining methods would be avoided in bonding this material?

18. What joining methods would you select for bonding thick PVC, which is heat-sensitive?

19. What is the difference between a tubular and a semitubular rivet?

20. What is the purpose of a blind rivet?

21. Define a clinch allowance.

22. Why are thread-forming self-tapping screws not used in thermosets?

23. What is a speed nut? An anchor nut? An insert? An expansion fastener?

24. Why would you use a threaded insert in a thermoplastic such as PVC instead of a drilled and tapped hole as in the case of metals?

25. A plastic hockey stick is to be made by cementing the stick to the blade in a lap joint. The joint will be flexed and thus will tend to be peeled open. How will you protect this joint against peel?

17

finishing
and
decorating

Finishing methods for plastics usually employ abrasives and polishing compounds in a variety of processes. Hand filing may be used to clean up molded articles and for deburring.

[17.1.]
abrasive finishing

Abrasive grinding is a useful method of finishing thermosets, especially those containing fillers and reinforcement. The grinding of thermoplastics is more difficult. These have low melting temperatures and tend to load grinding wheels and abrasive belts. Wheels and belts with an open grain, that is, with voids between the grains to act as chip spaces, should be used, and a coolant is required. The finer grades of abrasive will load readily with softened thermoplastics. Very soft and deformable plastics such as polyethylene foam cannot be ground.

[17.2.]
buffing

Buffing of plastics is employed for the removal of small surface defects and to obtain a lustrous surface. Buffing wheels consist of muslin disks, either sewn together to provide a firm surface or unsewn for flexibility. The stitched wheels cut faster but generate more frictional heat. They are suited to the polishing of parts that do not have complex shapes. The loose buffs are used for more irregular shapes or for entering crevices. Being softer, they generate less heat. Both types of wheel are charged with a cutting compound such as tripoli.

Polishing wheels are similar to buffing wheels, except that soft flannel disks are employed. These are charged with wax compounds.

A buffing sequence may often include ashing, polishing, and wiping. The ashing wheel is dressed with a slurry of pumice and water. The plastic part is held lightly against the wheel and kept in motion to prevent overheating and uneven ashing. The wheel must be rotated slowly to retain the slurry. When ashing is completed, the part is rinsed and dried.

The polishing compound is applied to one-half the width of the polishing wheel. The plastic part is first held against the charged half of the wheel, then against the uncharged part of the wheel to wipe off the polishing compound.

In these buffing operations, excessive wheel speed and hard wheels must be avoided.

For wiping, a clean, soft, and uncharged buffing wheel is used. Compounds are sometimes applied to the wiping wheel to remove grease.

[17.3.]
barrel finishing

Barrel finishing is a process of tumbling parts slowly in a rotating barrel. In addition to the parts to be tumbled, the barrel is also charged with an abrasive or a polishing compound to accomplish whatever purpose the barrel finishing must serve. This process is used to remove flash, to smooth rough corners, to grind, or to polish.

Tumbling may be done wet or dry. Wet tumbling is used to remove material, as in the removal of flash or burrs, while dry tumbling supplies polish.

It is difficult to make general statements about this process, because there are so many variables such as speed of rotation, tumbling medium, weight of part, and size of part. Setting up a successful tumbling process is done by trial. The proper speed depends on the purpose of the tumbling, the size of the part, and its shape. Too low a speed does not produce enough friction between parts, whereas too high a speed may damage the parts by dropping and impact. The tumbling compounds include sawdust, waxes, abrasive particles, wood pegs, and many others.

[17.4.]
preparation for decorating

The use of a dry cloth for cleaning plastic articles generates an electrostatic charge that attracts dust to the plastic. Many plastics are best wiped with a sponge with lukewarm water and a household detergent. Drying may be done with a soft cloth.

Surface preparation of thermoplastics for painting usually requires only a solvent wipe. However, the solvent wipe may not be practical for polystyrene, which could be dissolved or stress-cracked by the solvent. The surface to be painted must be free of dust, mold releases, waxes, plasticizers, and fingerprints.

Many thermoplastics that are molded into furniture components, polystyrene and foamed polyurethane especially, require painting or staining in order to resemble wood. Many of the solvents used in wood finishing operations, such as naphthas and ketones, will dissolve thermoplastics, especially polystyrene; some may even attack polyurethane. To protect polystyrene against solvents in the finishing material, a barrier coat is needed. In addition to protection, the barrier coat must provide color and satisfactory adhesion for the top coats. A finishing treatment for a furniture item may require three coats: barrier coat, wiping stain or tones, and a lacquer to provide a hard finish. The lacquer dries at room temperature by solvent evaporation. Although enamels have excellent gloss and hardness, they require curing temperatures at which most thermoplastics will distort. Poured polyurethane foam, however, is sometimes painted by using the paint as a mold release.

The application of printing inks to PVC and PS films presents no serious problems, but polyethylene film requires a pretreatment for ink adhesion. Three methods are available: chemical, flame, and electrical. Chemical methods are inconvenient and thus are rarely used. Such methods include treatment of the surface with ozone, chlorine, and sulfuric-chromic acids. But electrical and flame methods are preferred.

In flaming polyethylene, the plastic surface is exposed to a gas flame for one to three seconds to oxidize the surface. Polyethylene bottles may be rotated against a flame or passed through a ring burner. In the electrical method a high-voltage, high-frequency source creates a corona or spark discharge to the film, thus oxidizing it.

[17.5.]
silk screen decorating

Silk screen painting is adaptable to simple or elaborate designs in one or more colors on flat or slightly curved surfaces. In the silk screen process paint or ink is forced through a stenciled silk, nylon, or stainless steel screen onto the plastic surface. The screen is a tightly woven fabric attached to a rectangular frame and masked so that the paint or ink is pressed through the screen only in the areas where the stencil is open. A rubber squeegee forces the paint through the screen (Fig. 17.1). By the use of multiple screens intricate and multicolor designs may be produced.

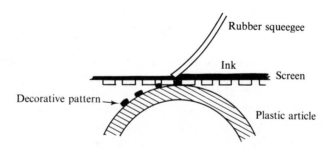

Rubber squeegee

Ink

Screen

Decorative pattern

Plastic article

FIGURE 17.1 — The silk screen method.

Acetals usually require an acid-etched surface before they will accept paint or print. Polycarbonate does not tolerate many solvents in paints and inks; most such solvents will etch PC or cause it to craze and lose its toughness. The cellulosics often contain plasticizers that migrate to the surface and soften paint films. The other thermoplastics are more tolerant of paints and ink formulations.

Thermosets are normally painted with the same baking enamels that are used with metals, since the thermosets can safely withstand normal oven-curing temperatures.

[17.6.]
electroplating and vacuum metalizing

Most of the plastics can be electroplated. The plastic surface is first degreased and dried, then made electrically conductive by depositing copper by chemical methods. The electroplated metal, usually chromium, is then deposited. Sometimes the electroplated part is given a final coat of lacquer. Parts to be electroplated should have no sharp corners or blind holes, because little metal is electrodeposited in such areas.

Vacuum metalizing deposits a thin film of metal, usually aluminum, on a plastic surface by vaporizing the metal in a vacuum. The metal film is only a few millionths of an inch thick. Adhesion of the metalized coat to the plastic is not as strong as in an electrodeposited coating.

The parts to be metalized are racked in a vacuum chamber. Staples of the metal to be deposited are hung on tungsten wires near the center of the chamber. After the required vacuum is pumped, the tungsten wires are heated by electric current, and the clips evaporate at relatively low temperature under the vacuum conditions. The vaporized metal condenses in a uniform and continuous coating on the plastic parts. A final protective coating of lacquer is required. Plasticized thermoplastics should not be metalized; the vacuum causes the plasticizer to migrate from the thermoplastic.

QUESTIONS

1. What methods are available for removing flash from molded parts?

2. For what applications are unstitched buffing wheels required?

3. What is meant by ashing?

4. Briefly describe how a plastic part is electroplated.

5. What is a barrier coat?

6. What is meant by an "open-grained" abrasive paper or wheel?

7. What is meant by "loading" an abrasive wheel or paper?

8. Why is too fine an abrasive paper ineffective when plastics are ground?

9. What is the principal purpose of wet tumbling?

10. What is the objection to wiping plastics with a dry cloth?

11. What methods are available for pretreating polyethylene to receive paint or inks?

12. Why are plasticized thermoplastics not vacuum-metalized?

index